Building Generative AI-Powered Apps

A Hands-on Guide
for Developers

Aarushi Kansal

Apress®

Building Generative AI-Powered Apps: A Hands-on Guide for Developers

Aarushi Kansal
Melbourne, Australia

ISBN-13 (pbk): 979-8-8688-0204-1   ISBN-13 (electronic): 979-8-8688-0205-8
https://doi.org/10.1007/979-8-8688-0205-8

Managing Director, Apress Media LLC: Welmoed Spahr
Acquisitions Editor: Celestin Suresh John
Development Editor: Laura Berendson
Editorial Assistant: Gryffin Winkler

Cover designed by eStudioCalamar

Cover image designed by imagii on pixabay (pixabay.com)

Distributed to the book trade worldwide by Springer Science+Business Media New York, 1 New York Plaza, Suite 4600, New York, NY 10004-1562, USA. Phone 1-800-SPRINGER, fax (201) 348-4505, e-mail orders-ny@springer-sbm.com, or visit www.springeronline.com. Apress Media, LLC is a California LLC and the sole member (owner) is Springer Science + Business Media Finance Inc (SSBM Finance Inc). SSBM Finance Inc is a **Delaware** corporation.

For information on translations, please e-mail booktranslations@springernature.com; for reprint, paperback, or audio rights, please e-mail bookpermissions@springernature.com.

Apress titles may be purchased in bulk for academic, corporate, or promotional use. eBook versions and licenses are also available for most titles. For more information, reference our Print and eBook Bulk Sales web page at http://www.apress.com/bulk-sales.

Any source code or other supplementary material referenced by the author in this book is available to readers on GitHub. For more detailed information, please visit https://www.apress.com/gp/services/source-code.

Paper in this product is recyclable

To my parents, for always backing my ambitious projects, even when they seemed a bit out there. This book is one of those dreams realized.

Table of Contents

About the Author

Aarushi Kansal is an experienced principal engineer. She has worked in a variety of technologies, including mobile development, Python, Go, and cloud, along with booming generative AI space. She has spearheaded AI initiatives in the workplace and regularly works on creative POCs in her spare time to stay at the top of the generative AI space.

About the Technical Reviewer

 **Akshay Kulkarni** is an AI and machine learning evangelist and a thought leader. He has consulted several Fortune 500 and global enterprises to drive AI and data science–led strategic transformations. He is a Google Developer Expert, author, and regular speaker at major AI and data science conferences (including Strata, O'Reilly AI Conf, and GIDS). He is a visiting faculty member at some of the top graduate institutes in India. In 2019, he was also featured as one of the top 40 under 40 Data Scientists in India. In his spare time, he enjoys reading, writing, coding, and building next-gen AI products.

## CHAPTER 1

Introduction to Generative AI

Generative AI (artificial intelligence) is a loaded phrase these days. Investors are throwing their money at it, execs are throwing it at each other, and at some point, a manager is probably going to ask you "can we do generative AI too?" or you're going to get tempted and hack together an LLM-powered bot at 2 a.m. This chapter introduces you, a software engineer, to the booming world of AI, by cutting through all the hype and demystifying AI. I start from the most popular architectures right now and then throughout the book to various models, both open and closed source. I aim to explain these models from the lens of a software engineer as opposed to a data scientist or machine learning scientist. This means the aim is to understand and explain just enough about the foundation models so you can customize and build AI-powered applications, leveraging these models. In particular I'll focus on diffusion models (you know all those cool AI images you've seen on your socials?) and transformer models (think ChatGPT, LLama, music-gen, etc.).

© Aarushi Kansal 2024
A. Kansal, *Building Generative AI-Powered Apps*,
https://doi.org/10.1007/979-8-8688-0205-8_1

What Is Generative AI?

Generative AI is essentially a kind of unsupervised or semi-unsupervised machine learning that allows people to create various types of rich content, like images, text, video, speech, and even music.

With unsupervised learning, a model is able to determine patterns in the data it is fed, often patterns the human eye would simply miss, without needing any kind of labelling. These models leverage neural networks (similar to the networks in our brains) to learn patterns and generate the rich content you've been seeing all over the Internet.

Semi-supervised learning is a combination of supervised and unsupervised learning. This means making use of a small number of labelled data (supervised learning) during the training or fine-tuning steps, combined with a large set of unlabeled data (unsupervised learning).

The ability to make use of unsupervised learning on massive amounts of unlabeled data (such as articles, books, images, etc.) is what supercharged companies' abilities to create massive foundational models such as GPT-4, Stable Diffusion, Llama Bark, etc. Without this style of machine learning, labelling what is essentially all of human knowledge (i.e., the Internet) would have been virtually impossible!

Okay, now that you have a high level intro into generative AI, let's talk a little bit about different architectures, in particular the two most popular: transformers and diffusion models.

Model Types

In this section, we'll explore two main types of architectures: transformers and diffusion models. While there are a range of architectures, I want to talk to you through the ones the foundation models used in this book are based on. Also, keep in mind, this section is not a deep dive, more of a summary, just enough so you know *what* you're using, when you build applications on top of these models. This means you won't be learning the math (but I do recommend you read the papers, research, and understand the math; it's fascinating!)

First up is transformers and then diffusion.

Transformers Explained

Transformers are currently dominating the natural language processing (NLP) space. Most of your favorite models are transformers, for example, GPT-4, Llama, Falcon, etc. Let's look into transformers and why this architecture becomes so popular. To do that, we need to go through a tiny history lesson.

Once upon a time, there were two main architectures: recurrent neural networks (RNNs) and Long Short-Term Memory (LSTM) networks (a type of RNN), specifically designed to handle sequential data (e.g., text). Let's discuss RNNs and then LSTMs.

RNNs

RNNs maintain a memory of previous inputs in their internal structure to process sequences of inputs.

Imagine reading a book and trying to predict the next word in a sentence. If you're reading word by word without remembering the previous context, it's tough. But if you recall the earlier part of the sentence, it becomes easier. RNNs do something similar: they remember the "history" to make sense of the current input.

Let's take a quick look at the basic workflow of an RNN in Figure 1-1.

1. **Input**: At each step, the RNN takes in an input (e.g., a word in a sentence). **(Xn)**

2. **Hidden State Update**: This input, combined with the previous hidden state **(hn)** (memory), is used to update the hidden state. This new state might carry forward crucial information and forget irrelevant details.

3. **Output**: Based on the updated hidden state, the RNN might produce an output (e.g., predicting the next word in a sequence). **The output is a combination of Xn and hn**.

4. **Move to Next Step**: The process repeats for the next element in the sequence.

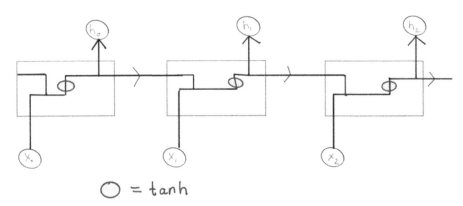

Figure 1-1. *RNN architecture*

While these basic RNNs are excellent for modelling sequential data like text or time series data, they have the fatal flaw of struggling to remember distant past information. In other words, they have a short-term memory. This tendency to forget is called the vanishing gradient problem.

This brings us to LSTMs, designed specifically for long-term memory.

LSTMs

The short-term memory problem is addressed by using LSTMs, which have a more complicated structure but function more similarly to how a human might read a book or hold a conversation.

In the previous RNN, you can see that the network is able to remember previous information because we pass the previous hidden state (h) into the current cell. Continuing on from this observation, you can also see how hidden states from further back cells become diluted; essentially the information in those states vanishes.

One of the core observations in Figure 1-2 is the top horizontal line, which transfers the vector straight through the cell and through the entire network. This means that information can flow through the sequence, essentially unchanged, meaning this network has the capability to remember information from further behind in the sequence. Kind of like a sushi train, food keeps passing along, and you can remove, modify, or leave the sushi as is.

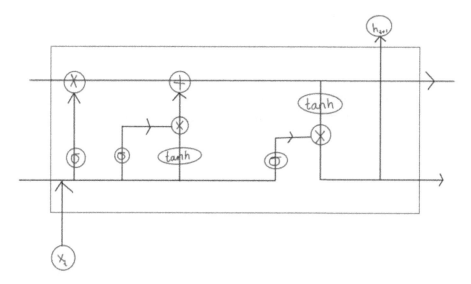

Figure 1-2. *LSTM cell*

But you also don't want to just pass information along with no modifications. The way that humans understand and process information is based on our ability to place more or less emphasis on different parts of a sentence or paragraph, based on context or prior knowledge.

To reproduce this kind of behavior, LSTMs use gates (forget, input, and output specifically) to determine what action to take.

So the basic workflow goes like this:

1. **Input Vector:** Similar to RNNs, the LSTM unit takes in an input vector and the previous hidden state at each time step.

2. **Gates in Action:**

   - The **forget gate** decides which parts of the cell state to throw away.

   - The **input gate** decides which values to update in the cell state.

   - After these updates, you have the new cell state that carries long-term memory.

   - The **output gate** determines what the next hidden state (short-term memory) should be.

3. **Output:** The LSTM produces an output, which is the hidden state passed to the next LSTM unit in the sequence.

4. **Move to Next Step:** The updated cell state and hidden state are passed to the next LSTM unit in the sequence, and the process repeats.

So with this variation of an RNN, you get an *improvement* on the vanishing gradient problem, but it's still not entirely solved. LSTMs remember for longer but not quite long enough.

Transformers

Fast-forward to 2017, a groundbreaking paper named "Attention Is All You Need" was published, with the key creation of a self-attention mechanism.

These models are able to track relationships between words and concepts and understand "context" in language – kind of like we as humans do instinctively, without even actively having to think about it. When humans talk about context, what we mean is attention. For example, when you're translating a piece of text from English to Spanish, you'll likely need to pay attention to words, not just next to each other, but distant from each other, because they can change the meaning, the tense, conjugation, and overall form of a word. Attention in the context of transformers is very similar. In other words, ensuring a neural network is able to glean context, because context heavily influences words in almost all NLP tasks.

Let's take a look at the overall workflow shown in Figure 1-3.

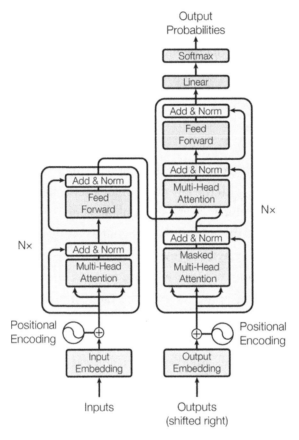

Figure 1-3. *Transformer architecture*
Image source: $https://arxiv.org/pdf/1706.03762.pdf$

1. **Input Representation**

   Tokenization and Embedding

   - First, your raw input (some text) is tokenized. This
     means breaking down the input into chunks – these
     could be words or characters.

- Then each token is mapped to a vector using an embedding layer. This vector representation captures the semantic meaning of the token. This is essentially representing words + meanings in numerical form.

Positional Encoding

- Since the transformer doesn't process data in sequence like RNNs, it doesn't have an understanding of order of tokens. So a positional encoding is added to the embeddings. This makes sure the model can account for the position of words in a sequence.

2. **Transformer Layers**

The core of the transformer model consists of a stack of identical layers. Each layer has two main components:

- Multi-head Self-Attention Mechanism

- Feed-Forward Neural Network

Multi-head Self-Attention

- This mechanism allows the model to focus on different parts of the input sequence when producing an output for a particular token.

- The "multi-head" part means this attention process happens in parallel multiple times, allowing the model to focus on different semantic aspects simultaneously.

- The attention mechanism uses three weight matrices called Query, Key, and Value, which help in determining how much attention to pay to various parts in the input sequence.

Feed-Forward Neural Network

- Each attention output is passed through a Feed-Forward Neural Network (separately but in parallel). The same network is applied to each position.

Residual Connection and Normalization

- After both the attention and feed-forward stages, there's a residual connection that helps in training deeper networks.

- Layer normalization is also applied after adding the residual connection.

- The residual connection helps with the vanishing gradient problem, and layer normalization aids in faster and more stable convergence.

3. **Output**

- If you're using just the encoder part (like BERT), the output is typically a vector representation of the entire sequence or individual tokens, which can be used for tasks like classification.

- If you're using the decoder part (like GPT-4), the output is another sequence, which is the result of transforming the input sequence.

4. **Additional Components**

Masking

- In certain situations, like training a language model, you don't want certain words to pay attention to future words in the sequence (because they shouldn't be "known" yet). Masking makes sure the model is "blind" to these future tokens during training.

- This is crucial for training models like BERT, where you want to predict a masked-out word without "cheating" and looking at it. For GPT, the masking makes sure that when predicting a token, the model can't look at future tokens.

Final Linear and Softmax Layer (For Tasks Like Language Modelling)

- The decoder's output can be passed through a final linear layer followed by a softmax to produce probabilities over the vocabulary. The token with the highest probability is usually taken as the prediction, especially for text generation.

- This is especially common in language modelling tasks where the goal is to predict the next word in a sequence (think ChatGPT).

So far, you've learned about generative AI in the context of language, that is, large language models (LLMs); next up is diffusion models, which have gained popularity in the image generation space.

Diffusion Explained

Most recently, diffusion models have been used by the likes of OpenAI for DALL-E, Midjourney, and Stability AI, all for image generation. The way diffusion models work overall is actually quite simple – one of the less complex concepts we'll discuss in this book.

Diffusion models are a type of generative model, which is used in a range of situations. You might already be very familiar with diffusion models being used for image and video generation. These models have also started showing promise in other areas such as drug discovery!

In Figure 1-4, you can see just all the places diffusion models fit in.

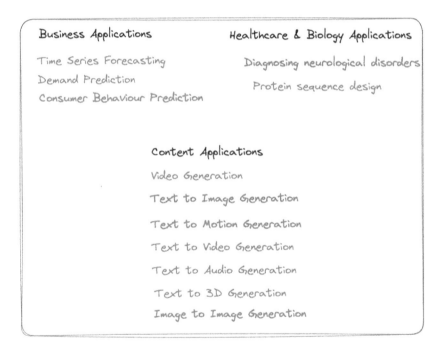

Figure 1-4. *Diffusion model applications*

We'll focus on images for the purpose of this book. Let's take a look at how these models work.

The Core Idea

Imagine a drop of ink spreading out in a glass of water. This process of diffusion, where particles move from regions of high concentration to low concentration, is a natural phenomenon. In diffusion models, a reverse process is used: it starts from a target data point (like an image) and gradually adds noise to it until it becomes a simple distribution. The magic is that this process can be reversed to generate new data samples.

How Diffusion Models Work

1. **Noise Addition Process (Forward Process) (shown in Figure 1-5)**

   - Starts with a real data sample (e.g., a real image).

   - Gradually, it adds noise over several steps until the sample becomes indistinguishable from pure noise.

2. **Noise Removal Process (Reverse Process/ Generation) (shown in Figure 1-5)**

   - Starts with a sample from a simple distribution (like Gaussian noise)

   - Uses a neural network to gradually remove the noise over several steps, guiding the sample to resemble a real data point from the target distribution

3. **Training**

- During training, the model learns to reverse the noise addition process. It gets better at transforming a noisy sample into a realistic one.

- This is done by using a neural network that predicts how to denoise a sample at each step. The model is trained on pairs of noisy samples and their less noisy versions.

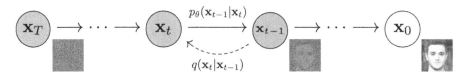

Figure 1-5. *Diffusion model process*
Image source: $https://arxiv.org/pdf/2006.11239.pdf$

And there you have it, you now know how Stability AI, DALL-E, etc., work under the hood.

What's Next?

So far you've learned about the architecture of the newly dubbed "foundation models." As you read through this book, the next topics focused on are LangChain (your Swiss Army knife to AI apps), monitoring (can you really go into production without solid monitoring?), and finally fine-tuning these models. As AI summer progresses, it's likely, rather than building and training models from scratch, you'll be fine-tuning foundational models to your needs.

Summary

This chapter has given you an understanding of the two most popular architectures out there: transformers and diffusion models. These are foundational models that will form the basis of the AI applications you build in future. Both open source and closed source models make use of these architectures. As an AI engineer (practicing or aspiring), it's important to understand what's going on under the hood of the tools you use.

CHAPTER 2

LangChain: Your Swiss Army Knife

In Chapter 1, you were introduced to the various types of generative models available, the most popular architectures, and how they work. Now, this chapter introduces you to LangChain, your Swiss Army knife to building robust applications on top of LLMs and other models. As you build applications beyond just making API calls, you're going to need various components to connect a model to your own data, to external data, and services, and that's what LangChain helps you with. A standard, modular way for you to essentially plug and play with models and various integrations.

As you go through this chapter, you'll be introduced to a few concepts you might not be intimately familiar with – don't worry, as I go along, I'll explain these concepts, and as you go through further chapters, you'll use these concepts in increasingly complex ways – which will help you further understand. Basically, this is the approach:

- – Introduce the concept and theory.
- – Learn by getting your hands dirty.

© Aarushi Kansal 2024
A. Kansal, *Building Generative AI-Powered Apps*,
https://doi.org/10.1007/979-8-8688-0205-8_2

The Whats and Whys

LangChain is not only one of my favorite frameworks for building AI-powered applications but also quickly becoming an industry standard. This framework provides engineers with a modularized, standard interface to plug different models (open and closed source), with various data sources and API integrations.

You can think of it like playing with an assorted bunch of building blocks to create almost anything you can imagine.

The main components LangChain is composed of are the following:

- LLMs
- Retrieval
- Memory
- Chains
- Tools
- Agents

By combining these concepts, you can create end-to-end LLM-powered apps that go beyond just a simple API call to OpenAI. You can chain calls, can allow your model to have access to various tools (e.g., Google search APIs), and finally, can use your LLM's reasoning abilities to decide which tools to use for particular tasks (this concept is agents).

In the next two chapters, you'll get to dive into each block or component. As I mentioned earlier, you're going to learn by doing, so the next few sections are broken down by use cases.

In the first use case, you're going to build an app to chat to your company or organization's Slack – you know, for when you have to look up certain information or messages? Instead of keyword searches and then scrolling through messages – why not chat to your archive? This use case will cover LLMs, retrieval, and memory.

In the next use case, you're going to build an agent that plans your day for you based on your mood, the weather, and your past preferences. This will cover chains, tools, and agents.

Let's move on to the first one, a chatbot.

Chatbot

From Chapter 1, you understand transformer models, and they're essentially predicting the next word/s. That's great for plenty of tasks like translation and text generation, but the thing that's really useful is the ability of an LLM to hold a conversation with you. That's the great thing about ChatGPT; it's been trained on *so* much data, and then on top of that, they've built a chatbot, so it's kind of like being able to chat with **everything**. You're going to learn how to build your own, smaller version for your personal data like notes, text messages, or Slack messages – without training or even fine-tuning.

Think about the ingredients in a human conversation for a second:

- All participants need to be able to speak the same language.

- Let's assume English for now, and we have that from most LLMs.

- The participants have to be able to remember what's happened in the past during this conversation.

- And access to knowledge in some way (in our case, knowledge of your Slack messages).

The last two points are referring to two concepts, which form the basis of *a lot* of LLM-powered applications you'll build:

- Memory

- Retrieval

Let's discuss both and then you'll start to build your chatbot.

Memory

With LLMs, by default, they have no concept of history or memory. Every query or call to an LLM is stateless, meaning they answer every question as if it's the first time it's been asked. And the model doesn't take into account your past interactions.

And that's the role this concept of memory comes into play – aptly named, it's a way to give an LLM remembering capabilities so you can hold a conversation with the model.

At this point, maybe you've already started thinking about how to start giving any LLM a memory.

One way would be to simply capture each query + LLM response and send that back into your LLM on the next query. As you can see in Figure 2-1 – you make a call to the LLM, it responds, and you parse the response, format it, and then send the response + your next question back as part of the context. You would keep repeating this pattern (until you run out of context length).

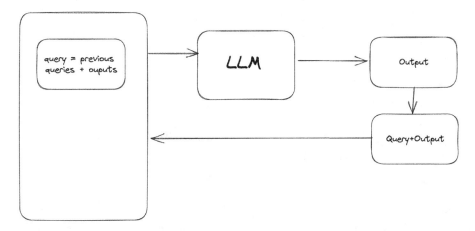

Figure 2-1. *Example of a simple way of just chaining queries and outputs for "memory"*

LangChain actually provides different chains to replicate this behavior in a reusable way – so you just have to decide on the type of memory you want to use, your input prompts, and some tunable parameters. Let's take a look with some real code.

Types of Memory

First, if you haven't already, go ahead and install LangChain (you can find installation instructions on the LangChain documentation page). For these examples, I'm going to use OpenAI, but you can use any other LLM – LangChain has integrations with the vast majority out there.

LangChain provides a number of types of memories for you to leverage; we'll focus on the four basic ones in this chapter and some of the more complex ones in later chapters.

First up is ConversationBufferMemory, which serves as a flexible memory buffer for chat conversations. It allows you to access the chat history in two formats:

1. As a string (buffer_as_str)

2. As a list of message objects (buffer_as_messages)

The class provides a load_memory_variables method that returns the chat history based on a chosen format. This output can then be used as context to your LLM, thereby providing it info on the previous parts of the conversation.

Let's take a look at a small example of how memory is represented:

```
from langchain.memory import ConversationBufferMemory
from langchain.llms import OpenAI

llm = OpenAI()
memory = ConversationBufferMemory()
memory.save_context({"input": "What is the capital of the UK"},
{"output": "London"})

print(memory.load_memory_variables({}))
```

Output:

```
{'history': 'Human: What is the capital of the UK\nAI: London'}
```

Here what you're doing is adding some history into your memory buffer. Right now, nothing is being passed into an LLM, but you can see the output of what would be passed to the LLM.

The prompt would include the "history," the "Human," and "AI" conversation – thereby giving the LLM context into the conversation.

In other words, ConversationBufferMemory is a simple way of representing historical context as a string that can be parsed and passed into a prompt.

Notice that the ConversationBufferMemory automatically formatted your input and output into the format of Human and AI conversation. This is the default, but you can change it using these variables:

```
human_prefix and ai_prefix.
```

For example:

```
memory = ConversationBufferMemory(human_prefix="Aarushi", ai_
prefix="Hermione")
memory.save_context({"input": "What is the capital of the UK"},
{"output": "London"})
```

Output:

{'history': 'Aarushi: What is the capital of the UK\nHermione: London'}

How might this look with an LLM attached?

1) Format the output so the LLM understands what this whole "history" thing is.

2) Pass that as a prompt + your next query.

3) Parse the response into your "history."

4) Rinse and repeat.

This process is shown in Figure 2-2

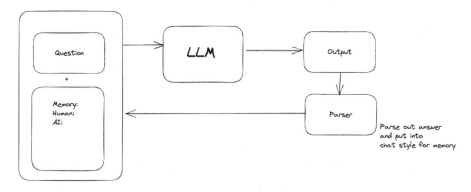

Figure 2-2. *Overall architecture of using memory in the prompt of an LLM*

Luckily, LangChain also has a built-in chain for just that. I'll go into chains later on, but let's take a quick look at how that would work.

```
from langchain.llms import OpenAI
from langchain.chains import ConversationChain

llm = OpenAI(temperature=0)
conversation = ConversationChain(
    llm=llm,
    verbose=True,
    memory=ConversationBufferMemory(human_prefix="Aarushi",
    ai_prefix="Hermione")
)

conversation.predict(input="What is the largest city in the UK
by population?")
conversation.predict(input="And second?")
conversation.predict(input="What about in Germany?")
# just to run one more time
conversation.predict(input="")
```

Okay, so similar to before, you initialize your ConversationBufferMemory, with the prefix you want (omit for defaults). Then all you do is add your questions – take note here, I've specifically kept the second and third questions brief with minimal context so you can see how it gleans context from the conversation memory.

When you run this, you should get a final output similar to the following:

```
The following is a friendly conversation between a human and an
AI. The AI is talkative and provides lots of specific details
from its context. If the AI does not know the answer to a
question, it truthfully says it does not know.
```

Current conversation:
Aarushi: What is the largest city in the UK by population?
Hermione: The largest city in the UK by population is London,
with a population of 8.9 million people.
Aarushi: And second?
Hermione: The second largest city in the UK by population is
Birmingham, with a population of 1.1 million people.
Aarushi: What about in Germany?
Hermione: The largest city in Germany by population is Berlin,
with a population of 3.7 million people. The second largest
city in Germany by population is Hamburg, with a population of
1.8 million people.

Basically, this chain abstracts away all the logic of parsing and updating the prompt from you so you just choose the memory type and related configuration.

As you use LLMs more and more, you'll start to notice two things:

1) Generally, LLMs have a maximum context length – meaning you can only really send a prompt of a certain size.

2) The larger your history or prompt, LLMs tend to start ignoring or missing older pieces of information.

So really, you want to send less of your history or maybe a condensed version of it – and once again, LangChain allows you to do just that with more types of memory, specifically ConversationBufferWindowMemory, ConversationSummaryMemory, ConversationSummaryBufferMemory, and ConversationTokenBufferMemory.

ConversationBufferWindowMemory – It is a variant of ConversationBuffer; it also keeps a history of your interactions, but only up to k number. This is a number you can decide on for your own needs – in my experience, I have found that for most use cases, a larger number is actually detrimental, and the model ends up hallucinating more often than not. I would recommend you experiment and find a balance of short but informational queries combined with a smaller window (k).

ConversationSummaryMemory – As the name suggests, this is a type of memory that condenses down your conversation into a summary that can be passed into your LLM.

Let's take a quick look at some code:

```
conversation_with_summary = ConversationChain(
    llm=llm,
    memory=ConversationSummaryMemory(llm=llm, human_
    prefix="Aarushi", ai_prefix="Hermione"),
    verbose=True
)
conversation_with_summary.predict(input="How are you
Hermione?")
conversation_with_summary.predict(input="What is the third
planet from the sun?")
conversation_with_summary.predict(input="second?")
conversation_with_summary.predict(input="fifth?")
conversation_with_summary.predict(input="")
```

Output:

The following is a friendly conversation between a human and an AI. The AI is talkative and provides lots of specific details from its context. If the AI does not know the answer to a question, it truthfully says it does not know.

Current conversation:

Aarushi asked how Hermione was doing and Hermione replied that she was doing great and asked how Aarushi was doing. Aarushi then asked what the third and fifth planets from the sun were, to which Hermione replied that the third was Earth, the only planet known to have life and the only planet in our solar system with liquid water on its surface, and the fifth was Jupiter, the largest planet in our solar system made up mostly of hydrogen and helium with a strong magnetic field and home to the Great Red Spot.

Here you'll notice instead of the conversation, with each participant, it's a summary of it – this kind of memory is great for particularly long conversations, especially where you need the LLM to understand the overall gist rather than each individual nuance of the conversation. For example, if you were chatting over internal HR documents, looking for general time-off policies – the LLM just needs the summary and the large points, not each bit of detailed policy.

Next up is ConversationSummaryBufferMemory, which combines both summaries and a buffer – meaning instead of only keeping a summary of previous interactions, it keeps interactions in a buffer as well as a summary. It means it keeps more recent interactions in a buffer and older ones as a summary (once the buffer hits a certain token length that you can tune).

Finally, ConversationTokenBufferMemory is similar to ConversationBufferWindowMemory but instead maintains a buffer of x tokens length rather than x number of interactions length.

So now we've covered some basic types of memory (there's more, but I want to save those for later chapters). At this point, with even just the simple code snippets shown previously, you have yourself a chatbot that remembers previous interactions and can hold a conversation with you based just on the knowledge it's been trained on, which is a lot, but it's not going to be your personal or company data (unless it's public). And we want to build a chatbot on your *own* information.

To do this, you could fine-tune your own model, and there are definitely use cases and reasons to fine-tune a model. But in our use case, that can be expensive and time consuming, and most importantly your own info such as Slack messages is going to change way too fast for you to be able to fine-tune fast enough. Think about the speed at which we message each other on Slack or any other messaging app. Luckily, there's been a rise of a new standard practice, called Retrieval Augmented Generation (RAG), to help give LLMs more knowledge without fine-tuning.

Retrieval

Retrieval is just one way of giving an LLM more specific, niche knowledge. It involves fetching data from some external source and passing it into your chosen LLM. Retrieval can be done any way – such as making an API call, reading a static file, reading a SQL DB, etc. This end-to-end flow of fetching info, passing it into an LLM, and the LLM generating a response is known in the industry as Retrieval Augmented Generation (RAG).

Diving into RAG

Since we're generally dealing with natural language and often unstructured messy data, the most popular (for good reason) storage system is a vector store. This essentially involves taking all of your niche data, creating a vector embedding, and storing in a vector database of your choice, as shown in Figure 2-3.

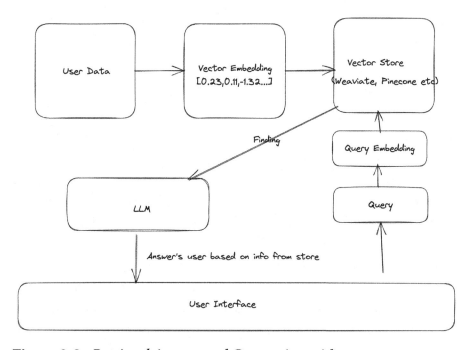

Figure 2-3. *Retrieval Augmented Generation with a vector store*

Let's walk through each component in detail, starting with embeddings.

Embeddings Explained

In layman's terms, embeddings are a way to represent everything that's not numbers (e.g., text, audio, images, etc.) as numbers. All ML models are basically math equations in some form, complex math but still just math – and so they don't actually understand or perceive words or images or anything else the way humans can with our five senses; they only understand numbers. That's why to deal with language, we need to convert words and sentences into a numerical representation that ML models can understand. The term "embeddings" is a general term for taking one type of data and representing it in numbers. Embeddings come in different types, such as graph embeddings, tensor embeddings, and many more.

In the context of LLMs, we're talking about vector embeddings, and you'll see vector embeddings and embeddings used interchangeably.

Vector Embeddings

As the name suggests, vector embeddings are a specific type of embedding where the representation is in the form of a vector. This means that the data, regardless of its original form, is translated into a fixed-length list of numbers.

Vector embeddings are commonly used in natural language processing (like Word2Vec or GloVe), where words or phrases are represented as vectors.

One of the major benefits of vectors is the ability to have vectors in high-dimensional spaces, which means that a vast number of features or aspects of language can be captured. Each dimension can potentially represent some facet of meaning, allowing the vector to encapsulate a rich set of semantic information.

And because vectors are basically long lists of numbers, we can do mathematical computations on data that normally wouldn't be possible. For example:

$$\text{vector(Germany) - vector(Berlin) + vector(France) =}$$
$$\text{vector(Paris)}$$

This shows that the difference between a country and its capital can be consistently represented in the vector space. So by knowing the capital of Germany and applying this relationship to France, we can deduce the capital of France.

Because we can do computations like this, another benefit of vector embeddings now is that we can use these geometric relationships between vectors to model semantic or functional relationships. For instance, in word embeddings, the vector difference between "dolphin" and "ocean" might be similar to the difference between "camel" and "desert," reflecting habitat relationships.

Basically, we use vector embeddings as a way to model intricate and complex semantic meaning and relationships between words and text.

Now, modelling these semantic relationships isn't a trivial task – you need an embedding model that understands these complex relationships and can take your raw words or sentences and create vector embeddings. Luckily there are quite a few embedding models available:

- OpenAI's embedding model, for example, text-embedding-ada-002

- Cohere's embedding model

- Open source embedding models (https://huggingface.co/BAAI/bge-large-en)

- And a lot more

LangChain also integrates with the vast majority, and you can have a look here: https://python.langchain.com/docs/integrations/text_embedding/.

So in an RAG application, you'll have one step that involves taking your raw data, inputting it into an embedding model, and then getting a vector embedding out of it, as shown in Figure 2-4.

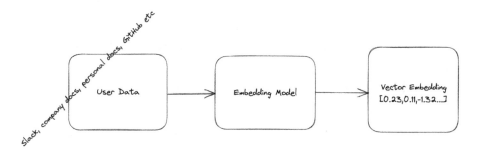

Figure 2-4. *Steps for taking unstructured, raw data and converting to an embedding*

Now that you have these embeddings – you need somewhere to store them, somewhere to search from so your LLM has an external data source. And that's where vector stores come into play.

Vector Stores Explained

Vector stores or vector DBs have been around long before generative AI became so mainstream – they were used in areas such as information retrieval, recommendation systems, and even molecular biology.

Now with modern embedding models and the rise of LLMs, there's also been a rise not just in popularity of vector stores but also new, more modern DBs available specifically designed to fit in with modern generative AI models.

Here's a non-exhaustive list for you:

- Weaviate

- Pinecone

- Chroma

- Qdrant

- Traditional DBs that have started supporting vector embeddings

If you want a more detailed list, check out `https://python.langchain.com/docs/integrations/vectorstores`.

So how do these DBs actually work and what's so special about them compared to existing SQL and NoSQL DBs?

Vector DBs are specifically designed to store high dimensional data like embeddings and allow for fast querying and lookups. They have the capabilities of traditional databases, while being optimized to handle the complexity of vector embeddings.

When you create + store an embedding, a reference to the original data is also stored. Then, when you make a query to the DB, the query is first converted to an embedding (using the same model), and this embedding is used to find the most similar content and return it to you.

Steps are shown in Figure 2-5.

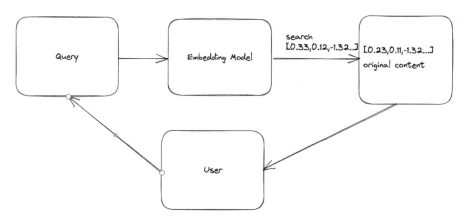

Figure 2-5. *Steps for converting a user query into a searchable query in a vector DB and returning the answer*

Unlike traditional databases that are optimized to search for exact values – a number, a string, or some other single dimensional, exact value – vector databases are optimized to search for vectors (high dimensional data) that are most *similar*, not exact to another query vector.

To do this, vector databases store your data in structures that allow for fast querying – called indexes. These indexes are created using a variety of algorithms, which we won't go into detail but are listed if you want to read more about them:

- Random Projection

- Product Quantization

- Locality-Sensitive Hashing

- Hierarchical Navigable Small World

When the query comes in, these DBs make use of various algorithms to do an Approximate Nearest Neighbor (ANN) search to get the most similar matches based on some similarity metrics, such as cosine similarity, dot product, Euclidean distance, or Hamming distance.

And there you have it, theory on how you get an LLM hooked up to your own data.

What Else Is RAG Good For?

Hallucination – that's probably a word you've heard *a lot* when discussing or critiquing LLMs or LLM systems. Hallucination basically refers to the LLM's tendency to just "make up" information. This can happen for reasons such as not having the information and being trained on incorrect information (remember, these systems are trained on public data, and there's a lot of misinformation out there).

By using RAG, you can help reduce this tendency by having your system fetch the correct information and including it in the context for when your LLM crafts a response.

Let's move on to some actual code.

The App

Okay, so now we're going to put all of these pieces together and build a chatbot over your Slack messages (or any other data source really).

Prerequisites

- Python 3

- Latest LangChain version

- A vector database

    - I'm going to use Weaviate's 14-day free hosted one, but you can choose one of your choice. The code generally remains the same.

- Slack messages

    - I've created a new workspace with some "fake"
      messages – you can do the same or if you are able to
      use real ones, use those.

- An LLM

    - I'm using OpenAI's GPT-4.

Loading Your Data

Since we're dealing with unstructured data here, the first thing you're
going to need to do is load your Slack messages in a form that can then be
turned into vector embeddings.

Go ahead and export your Slack messages as a zip file from
{your_slack_domain}.slack.com/services/export.

Once you have that, let's load it as seen in the following code snippet:

```
LOCAL_ZIPFILE = "gen_ai_co_slack.zip"  # Paste the local paty
to your Slack zip file here.
loader = SlackDirectoryLoader(LOCAL_ZIPFILE)
docs = loader.load()
```

Basically, you'll most likely be dealing with unstructured data, so
LangChain provides you with a lot of different types of "loaders" that
take data of one type (e.g., Slack, epub, logs from Datadog, Excel, GitHub,
and many more) and turn them into structured data. For example, the
SlackDirectoryLoader takes the json files exported from Slack and converts
them into a list of documents. This Document structure just stores text and
its associated metadata.

For example, the json files end up looking something like this:

```
Document(page_content='<@U05SQ9E71EF> has joined the
channel', metadata={'source': 'q4-planning - U05SQ9E71EF -
1694993629.080429', 'channel': 'q4-planning', 'timestamp':
'1694993629.080429', 'user': 'U05SQ9E71EF'})
```

where page_content is the messages and the other fields are associated metadata. Other loaders work in different ways – but the end result is always the same, unstructured data converted into structured.

Transforming Your Structured Data

Now, you have your data loaded, but you still need to transform it before you can create the embeddings – this means transforming your data into smaller chunks before creating an embedding. This is because you want to be able to fit meaningful parts of your data within the context window of your model when querying and adding answers as context. This is where LangChain has document transformers that you're going to use. The default you're using here is the RecursiveCharacterTextSplitter, which tries to split on certain characters – by default on \n\n, \, " ", and "".

In this snippet, I've chosen a chunk size of 500 and an overlap of 40 – the overlap ensures continuity between chunks.

```
text_splitter = RecursiveCharacterTextSplitter(chunk_size=500,
chunk_overlap=40)
documents = text_splitter.split_documents(docs)
```

You can check out the other document transformers here: https://python.langchain.com/docs/integrations/document_transformers/.

Now, choosing a chunk size can almost be a bit of an art form. Remember, the size of your chunk influences your embeddings.

If you go with shorter chunks (think words, or sentences), then your embeddings will lose the wider context in the paragraph – the embedding will narrow down on the specific meaning of the word or sentence.

If you go with longer chunks – you'll get the broader context, but that could add confusion and actually cause the embedding to lose the more specific or nuanced meanings you might need.

As such, you need to really take into account:

What kind of documents are you dealing with? For example, Slack messages are usually quite short, so you can easily go for a shorter chunk size. Books and scientific articles are a different story – they're longer and often you need the wider context; in this case, I would consider the next questions.

What is your use case? Will you or your users be asking very specific, nuanced questions? Short or long queries? Vague queries? For example, if your application is more of a very specific Q&A application, I would go through the documents and get a feel of how long do I as a human being need to read to get the right answer and based on that choose my chunk size.

A lot of these questions can be answered through you experimenting with sizes.

Embeddings and Storage

Next is creating your actual embeddings out of these chunks and actually storing them somewhere. Again, LangChain provides an abstraction layer to various embedding models and vector stores. This means you can just plug in any one you have access to.

In the next example, I'm going to use OpenAI for embeddings and Weaviate for storage, but since it's a plug-and-play concept, you can replace it with one of your choices and the overall code doesn't need to change drastically.

Check out all the embedding integrations here: `https://python.langchain.com/docs/integrations/text_embedding/openai`.

```
embeddings = OpenAIEmbeddings()
db = Weaviate.from_documents(documents, embeddings, weaviate_
url=WEAVIATE_URL, by_text=False)
```

And your code is as simple as this; choose the embedding integration and the vector store and pass your chunked docs and embedding model and you get a vector store populated with your embeddings, that you can now query.

For example, in my setup, I can run this query:

```
query = "What is the work from anywhere policy?"
docs = db.similarity_search(query)
```

and get a response like this:

Exciting News! We're officially launching our *Work from Anywhere (WFA)* policy. Starting next month, you'll have the flexibility to choose your work location, be it from home, a café, or any place that boosts your productivity.

So you can see my query was turned into an embedding; a similarity search took place; the resultant content was returned to me as is.

Okay so, maybe now you're wondering how to choose an embedding model.

Here are some of my considerations:

– Cost: Hosted ones like OpenAI can be expensive.

– Latency: Hosted ones are quite new currently and often don't provide SLAs, so expect unexpected latencies.

- Quality: This one's tricky because it's unlikely you're going to be able to test out *all* the models closed and open source out there. I use the Massive Text Embedding Benchmark (MTEB) leaderboard as a good source (`https://huggingface.co/spaces/mteb/leaderboard`).

Memory

And finally, now that you can retrieve data, you also want to hook up a memory component.

By now, you know how to instantiate memory, so go ahead and do that. The new step you're going to do is include a so-called "retriever," which is just the vector store from which your app can fetch. It's a VectorStoreRetriever object, which has functions on it to allow it to actually query the store. Instantiate it as shown here:

```
llm = OpenAI(temperature=0)
memory = ConversationSummaryMemory(llm=llm, memory_key="chat_
history", return_messages=True)
ret = db.as_retriever()
```

Now, previously we used a ConversationChain, in this case, we're going to use a different chain that can handle a retriever, called a ConversationalRetrievalChain.

This chain is similar to the chain you've used previously, except it includes one extra step internally, when you ask questions. It actually passes your question directly to the vector store and returns the stored documents.

Essentially, it's an abstraction on this call:

```
docs = db.similarity_search(query)
```

In the following code snippet, you'll see how to set up your chain – now when you run this, you'll see the same memory + summarization combination you saw previously.

```
qa = ConversationalRetrievalChain.from_llm(llm, retriever=ret,
memory=memory)
qa("What is the work from anywhere policy?")
qa("are there any in office days required?")
qa("any coworking?")
```

And there you have it, a conversational app across your Slack messages.

What's Next?

Okay, so far you've built the seedlings of a conversational chatbot across your Slack messages. Next you're going to learn in depth about chains and agents. This will help take your app to the next step – moving from purely a Python script to something slightly more interactive.

Summary

In this chapter, you were introduced to LangChain and you learned how to use the two most basic building blocks: memory and retrieval, which allow you to create Retrieval Augmented Generation applications. This is a great start to giving your LLMs external knowledge, without having to spend time and effort on fine-tuning. On top of this, you can keep updating your vector store with new information, much faster than you could fine-tune with new information.

In the next chapter, you'll take what you've learned about memory and RAG one step further and create an agent using LangChain.

CHAPTER 3

Chains, Tools and Agents

In Chapter 2, you learned about RAG, memory, retrieval, and embeddings. You were able to combine these concepts and build yourself a command-line chatbot that answered your questions *and* could remember the rest of your conversation. This allowed the LLM to become "smarter" by getting context from history. Your chatbot also had access to up-to-date, personal information via a vector database, meaning it was able to answer questions beyond what it was trained on. This also helped prevent hallucination.

Now you're going to take it one step further and build an agent – an independent application that can access the world and make its own decisions on what steps to take to get to the final goal.

High-Level Concepts

Before going straight to code, I want to walk you through some theory on the concepts you'll be making use of. In particular, I'll talk you through chains, tools, and agents.

Chains

First concept (which you actually used briefly in Chapter 2) is a chain. These are wrappers around multiple various components – ranging from LLMs, APIs, libraries, databases, utility functions, etc. They are one of the

© Aarushi Kansal 2024
A. Kansal, *Building Generative AI-Powered Apps*,
https://doi.org/10.1007/979-8-8688-0205-8_3

core components of LangChain and enable you to really augment your LLM in a structured and easy way. You can craft your own chains or you can use the many existing ones. Chains are super important because they allow you to become a lot more creative with LLMs and solve increasingly complex problems, through integrating various entities.

These chains can be really simple such as just having one LLM or they can be increasingly complex – by combining multiple entities (also sometimes called utility chains).

The ones you've already used are related to RAG and conversational history. Recall in Chapter 2 you used the following:

- ConversationChain

- ConversationalRetrievalChain

The conversation chain you used, extended another, simpler chain, called an LLMChain, which alone, just receives a prompt and LLM and makes the call to the specified LLM and spits out the output. This is one of the most basic chains, and the ConversationChain builds on this algorithm to load historical context into the prompt that is then passed into the LLMChain and queried.

The next one you used was ConversationalRetrievalChain, and it is a chain specifically for retrieving information from a data source (in Chapter 2, that data source was Weaviate). This one is a little more complex, as it does three major things:

- It takes the chat history in and crafts an entirely new question based on history and new query.

- This question is passed into the retriever (i.e., this becomes the query to Weaviate).

- After getting the right documents, it passes the original question and fetched documents into the LLM to get a response.

These are just two chains; there are a lot more available for you to use – I recommend you check them out and build some on top of them for even more customized use cases.

Okay, so now you understand the concept of chains and glueing multiple utilities together.

Think of chains the same as a human body lifting an arm, or yawning, or lifting a mug and drinking from it – all one smooth action, but a lot of little things happening and interacting with each other under the hood to make them happen.

And if chains are larger actions, you can think of tools as something that enhances your abilities and/or knowledge, for example, the ability to do complex math or execute Python code.

So now let's talk about the tools that give your LLM further access to the world.

Tools

Tools are wrappers that allow your LLM to interact with the world. This is a fancy way of saying; these are essentially functions that take some sort of input and output something based on it.

For example, if you were using a search tool – your input might be a query like "best sushi restaurants in London," and the output you would get is a list of top sushi restaurants in London. This is information that could then be fed into your LLM and used further – maybe to transform that list into a "tour of London's best sushi places" or maybe make recommendations to your user based on their dietary needs.

These tools can be simple API calls, chains themselves, agents, or anything else that *does* something when given an input.

And again, LangChain provides you with a range of pre-built tools, for example:

- – Search tools
- – Bash script tools
- – YouTube tools
- – Weather tools
- – Python REPL

And many more – check out their documentation for the full list.

For the majority of use cases, a combination of these pre-built tools should get you where you need, but you can also craft your own tools if needed.

Building a Custom Tool

Building your own tool is fairly simple, and you can get as complex with it as you need.

Within the LangChain framework, you have the BaseTool class, which is your blueprint to building a tool.

The main components of this blueprint are as follows:

- – Name
- – Description
- – _run function – Default function that runs when tool is called
- – _arun function – Function if you want async running

The name and description are required fields, and there are a few more optional fields that you can check out in the library if interested.

Let's discuss the description field though – this is one of the most important fields because it's what your LLM uses to make the decision on when/how and why to use this tool.

Some best practices for you to consider:

- Clearly state when to use the tool.

- State how (especially if it's a more complicated tool).

- State when to *not* use the tool.

  - I've found this one very useful when using multiple tools inside of an agent. By being clear on when to not use the tool, you can really assist your LLM in becoming more accurate, as LLMs have the tendency to also just use a tool if they're not sure exactly which one to use or if there isn't one that best matches its need.

- Provide some examples of using the tool.

  - This one is great for helping your LLM reason by seeing.

Okay, let's see some code. This one is going to be a really simple tool that just reverses any string passed into it.

As you can see in the following, you have to extend the BaseTool class, provide a name and description, and implement the _run method. I have not implemented the async function – but you definitely can for your own use case, if needed.

```python
from langchain.tools import BaseTool
class StringReverseTool(BaseTool):
    name = "String Reversal Tool"
    description = "use this tool when you need to reverse
    a string"
def _run(self, word: str):
    return word[::-1]
def _arun(self, word: str):
    raise NotImplementedError("Async not supported by
    StringReverseTool")
```

Now that you understand chains and tools, I want to show you agents, which is where LLM development gets really interesting. Everything you've learned so far can be combined into one or multiple agents.

Agents

Agents are one of the most interesting, creative, and kind of buzzy concepts in the AI space currently. They are super powering and completely transforming the way we do complex tasks – previously only considered doable by human beings.

In more concrete terms, an agent is an application that is powered by an LLM and interacts with different APIs, entities, libraries, chains, tools, etc.

The LLM is the "brain" that makes the decisions on which chain and/or tools to execute, what to do with the output, and how to interpret various inputs/outputs and human interactions.

When ChatGPT first came out (as well as the other chatbots out there), the main way of interacting with GPT-3 was going on to OpenAI, writing some prompt, getting an answer, and then maybe continuing questioning or asking for different formatting, more in-depth answers, clarification, etc. This is a very manual and human process. In this process, the human is the decision maker, and the human has a goal or task to achieve.

Agents, on the other hand, attempt to replicate this human goal-oriented behavior – given a goal, they will determine their own tasks to achieve the tools to use and how to process the output of the tools and craft their own prompts to help them achieve each step to their final goal.

So what is this independent, self-thinking application actually made of?

1. An LLM

2. Memory

3. An agent

4. One or more tools

5. Agent executor

    a.   This is what runs the actual code when told
          to do so.

You can see the architecture in Figure 3-1

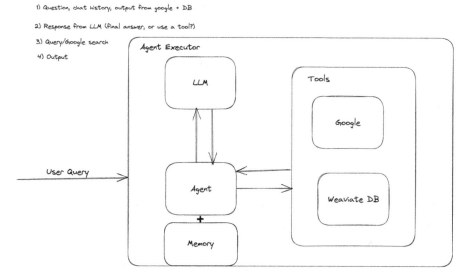

Figure 3-1. *Architecture for an agent – an independent, self-thinking application*

Let's dig a little into this component called an agent. At the crux of it, an agent is really a way of forcing the LLM to "think," that is, a way of prompting the LLM to think in a certain style. For example, a very simple way would be to just say "think step by step" after asking a question. And since the onset of AI summer, there have been numerous papers on various algorithms and styles of prompting LLMs to facilitate better logic and reasoning and minimize hallucination.

LangChain provides a range of these pre-built for you. Let's dive into some of the most commonly used ones:

- zero-shot-react-description
- react-docstore
- conversational-react-description
- chat-zero-shot-react-description
- chat-conversational-react-description
- self-ask-with-search

Here, maybe you've noticed a lot of them seem to have the word "react" in them. This is a fairly new framework for prompting LLMs. So let's talk about the basic premises of ReAct.

ReAct

ReAct stands for reason and act, and it's a framework that was proposed in March 2023 and has gained significant traction since then.

You can read the full paper here: `https://arxiv.org/pdf/2210.03629.pdf`.

The goal of ReAct is to create a train of reasoning along with actions based on that reasoning and interweaving the two – meaning reasoning something, based on that taking an action and then based on the actions output reasoning again and taking another action until the task is achieved.

This is shown in Figure 3-2. Here you can see the LLM reasons that it needs to search Cirque du Soleil, find the hotel, and then find the number of rooms in the hotel. It then takes different actions and "observes" the output and based on that reasons or thinks to itself again and makes another action until it finally comes up with an answer.

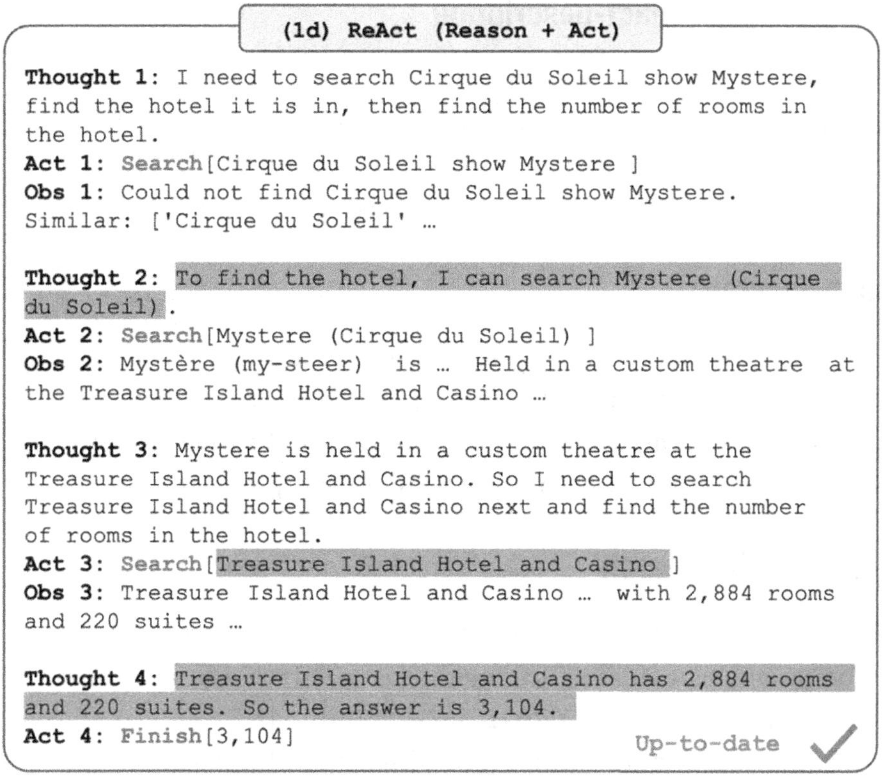

(1d) ReAct (Reason + Act)

Thought 1: I need to search Cirque du Soleil show Mystere, find the hotel it is in, then find the number of rooms in the hotel.
Act 1: Search[Cirque du Soleil show Mystere]
Obs 1: Could not find Cirque du Soleil show Mystere. Similar: ['Cirque du Soleil' …

Thought 2: To find the hotel, I can search Mystere (Cirque du Soleil).
Act 2: Search[Mystere (Cirque du Soleil)]
Obs 2: Mystère (my-steer) is … Held in a custom theatre at the Treasure Island Hotel and Casino …

Thought 3: Mystere is held in a custom theatre at the Treasure Island Hotel and Casino. So I need to search Treasure Island Hotel and Casino next and find the number of rooms in the hotel.
Act 3: Search[Treasure Island Hotel and Casino]
Obs 3: Treasure Island Hotel and Casino … with 2,884 rooms and 220 suites …

Thought 4: Treasure Island Hotel and Casino has 2,884 rooms and 220 suites. So the answer is 3,104.
Act 4: Finish[3,104]   Up-to-date ✓

Figure 3-2. *Reason + Act examples from the original ReAct paper*

And this concept is what most of the main agents are based on. The primary difference is what they are optimized for normally through tools, memory, and vector databases.

zero-shot-react-description

This agent type has no memory; it can only execute on one interaction. It will reason and output based on the ReAct framework but will not remember its previous thinking, or final answer on any subsequent uses.

conversational-react-description

This agent is the next enhancement on the zero shot agent – it allows for a memory. You can plug in any kind of memory. Recall in Chapter 2, you used ConversationBufferMemory and ConversationSummaryMemory; you could use these or you could also use an external storage as memory.

react-docstore

This agent uses ReAct but is optimized to use something called a Docstore in the context of LangChain. Basically an agent that uses some document store as a tool and can search in it for more context. The built-in ones include Wikipedia and In Memory (a Python dict representation).

self-ask-with-search

This agent is based on another proposed method to improve an LLM's reasoning and logic abilities. This method is called self-ask, and you can read the paper here: `https://ofir.io/self-ask.pdf`.

The concept is to get the model to ask itself a series of questions, answer those, and repeat until the final answer is reached.

This can involve giving an example of self-asking in the prompt.

This is shown in Figure 3-3.

Self-Ask

GPT-3

Question: Who lived longer, Theodor Haecker or Harry Vaughan Watkins?
Are follow up questions needed here: Yes.
Follow up: How old was Theodor Haecker when he died?
Intermediate answer: Theodor Haecker was 65 years old when he died.
Follow up: How old was Harry Vaughan Watkins when he died?
Intermediate answer: Harry Vaughan Watkins was 69 years old when he died.
So the final answer is: Harry Vaughan Watkins

Question: Who was president of the U.S. when superconductivity was discovered?
Are follow up questions needed here: Yes.
Follow up: When was superconductivity discovered?
Intermediate answer: Superconductivity was discovered in 1911.
Follow up: Who was president of the U.S. in 1911?
Intermediate answer: William Howard Taft.
So the final answer is: William Howard Taft.

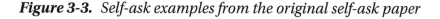

Figure 3-3. *Self-ask examples from the original self-ask paper*

In the paper, one of the enhancements on self-ask was to include a search engine. From the paper:

"self-ask clearly demarcates the beginning and end of every sub-question.

Therefore, we can use a search engine to answer the sub-questions instead of the LM. Search engines have features that LMs lack, such as an ability to be easily and quickly updated"

This just means a search tool allows the LLM to have access to more recent, up-to-date information as well as access to information retrieval algorithms and abilities under the search API/engine.

Okay, now that you've learned about agents, let's get to actually making one yourself.

The App

Okay so the app you're going to build is a day planner for any given city. It'll be able to take into account the weather, understand what kind of activities you want, and give you tailored recommendations.

For that, you'll need two tools to start with:

- Weather

  - Specifically, OpenWeather, but if you wanted to, you could also use another API and write a custom tool.

- Up-to-date info about places in a city

  - Specifically Google API, but again you can use another one

So with these tools in mind, take a look at Figure 3-4 for the overall setup of your agent. It's going to have memory and access to an array of tools, and agent executor will help orchestrate.

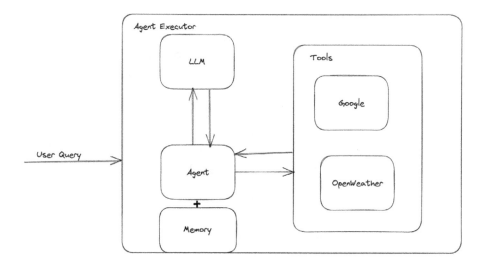

Figure 3-4. *Reason + Act examples from the original ReAct paper*

Okay, now the actual libraries I'm going to use:

- For the UI: Streamlit – There are others for you to try as well.

    - Gradio and Chainlit are the other two most popular ones.

- LangChain built-in tools:

    - OpenWeatherMapAPIWrapper

    - GoogleSerperAPIWrapper

        - This is a wrapper around serper.dev, which gives me access to Google search (and other APIs if desired).

        - LangChain also has wrapper for direct Google API access such as GoogleSearchAPIWrapper or GooglePlacesAPIWrapper.

- Agent Type:

    - I'm going to use a conversational-react-description agent, because I want both ReAct and a memory.

On to the code, I won't go through the entire code base; you can see that on GitHub, just the parts of note.

In the first code snippet, you're setting up your tools. You instantiate them and pass them into a Tool object, with a description and which function should be run. This is what tells the LLM what each tool can be used for – and allows the LLM to make the decisions. The agent executor uses whatever is in the func field, to actually execute, when the LLM makes a decision.

```
tools = [
   Tool(
       name="Search",
       func=search.run,
       description="Useful for when you need to get current, up
       to date answers."
   ),
   Tool(
       name="Weather",
       func=weather.run,
       description="Useful for when you need to get the current
       weather in a location."
   )
]
```

And then you set up the memory (recall, you did this in Chapter 2) as shown here:

```
memory = ConversationBufferMemory(memory_key="chat_history")
```

Then you set up an LLM chain; recall from the chain section, this chain is one of the most simple chains, and all it does is make the call to the LLM and get the output.

```
llm_chain = LLMChain(
    llm=ChatOpenAI(
        temperature=0.8, model_name="gpt-4"
    ),
    prompt=prompt,
)
```

Also, take note here, you can replace the LLM field with an LLM of your choice. I'm using the wrapper for GPT-4 – but LangChain has wrappers for many others.

And then you set up your agent; this is where you pass in any chains, tools, and memory. Take note here of max_iterations. I've set this to 3 because the ReAct framework technical could go on for almost an infinite number of loops for more curly queries. And even for less complex ones, there is a chance it could loop through many, many times, and since each loop costs money (i.e., a call to an API), I recommend locking down the number of iterations. Even for a self-hosted model, locking down iterations is a good idea depending on your use case; otherwise, the agent might take just way too long to come up with an answer for you.

```
agent = ConversationalAgent(llm_chain=llm_chain, tools=tools,
verbose=True, memory=memory, max_iterations=3)
```

Finally, you set up the agent executor that takes in the agent, tools, and optionally callbacks.

```
agent_chain = AgentExecutor.from_agent_and_tools(
    agent=agent, tools=tools, verbose=True, memory=memory
)
```

Now, if you go ahead and run your application with

```
streamlit run day_planner_agent.py
```

it'll take you to your nice UI, where you can start querying it.

First, I put in my request about Melbourne, food, hiking, and not liking rain. Notice how the response includes info on the weather and sushi places and hiking places.

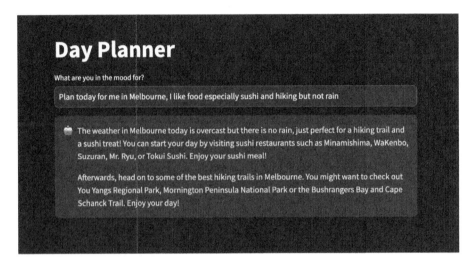

Figure 3-5. *UI and Input + Output for your new Day Planning Agent*

Then in Figure 3-6, you can see the exact ReAct framework being executed. The main concept being first a thought such as "Do I need to use a tool?" then an action either use a tool or no tool and get an answer. Then an observation based on the output of the action taken. Then a thought, then action, then observation, and so on, until you get a final answer.

Figure 3-6. *Reason + Act for your own agent*

And there you have it, you have an agent that can reason and access the outside world.

The next steps for you would be to take a look at LangChain and see what kind of agents you would like to build. In this example, it has been a chat interface, but you can decide on the kind you want, maybe no interface, maybe it just runs continuously in the cloud somewhere.

Summary

In this chapter, you built your first agent that had reasoning abilities and access to the external world. You learned the next building blocks in LangChain: chains, tools, and agents. With this knowledge, you can start building some more complicated agents to do certain tasks for you. Remember, everything in LangChain is plug and play, so try experimenting and plugging in new libraries and tools.

CHAPTER 4

Guardrails and AI: Building Safe + Controllable Apps

In Chapter 3, you combined all your learnings on RAG, memory, and embeddings with tools and chains to create an end-to-end agent – that could plan out your day for you. This agent was able to reason and have access to "the world" via API integrations (the so-called tools). This was a fairly simple application, but it was still autonomous – and when AI is autonomous, there's always space for things to go wrong if proper safeguards are not in place.

This chapter delves into the critical aspect of ensuring safety and reliability in AI-powered applications through the concept of "guardrails." Using NVIDIA's open source library, NeMo Guardrails, you will explore strategies to counter common challenges in conversational AI systems, such as hallucination, topic drift, and ineffective moderation.

© Aarushi Kansal 2024
A. Kansal, *Building Generative AI-Powered Apps*,
https://doi.org/10.1007/979-8-8688-0205-8_4

Why Guardrails?

Chatbots and conversational and generative AI have so many benefits but also a lot of pitfalls and dangers:

- AI hallucinates and convincingly makes up information.

- AIs can be very difficult to stay on topic.

- AIs don't inherently (as of yet) know how to effectively moderate or end conversations.

- LLMs can output toxic, hateful, and harmful information.

- LLMs can inadvertently leak data and PII, especially when malicious actors are on the other end.

You can effectively address these concerns to a certain extent through carefully constructing prompts and using techniques such as Retrieval Augmented Generation. So far though, this is not enough on its own – not if you want to release a generative AI app with minimal human intervention into the wild.

This is where the concept of "guardrails" comes into play – a way to provide structured, reliable guidance to your AI-powered application.

Think of guardrails as a set of rules or guidance to prevent your LLM or chatbot from acting poorly – as determined by you. It's the same as giving human employees a manual on how to behave and handle certain situations and what topics are and what aren't.

In Figure 4-1, you can see a representation of where guardrails roughly sit; they aim to protect your LLM from issues such as jailbreaks, hallucination, going off topic, and general moderation.

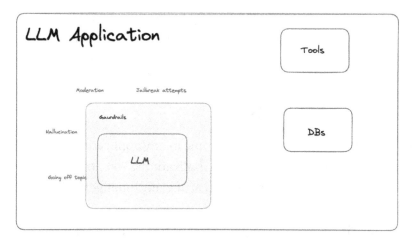

Figure 4-1. *Guardrails for concerns such as hallucination and moderation*

NeMo Guardrails

Enter NeMo Guardrails from NVIDIA – an open source library to add guardrails to your LLM-backed applications. At the time of writing it's in its alpha release, but definitely worth exploring and starting to adopt with care.

Firstly, let's talk about some of the possibilities with this library.

Keeping Your Bot on Topic

Okay, so one of the biggest pitfalls (*and* benefits) of the current set of LLMs is the fact that they're trained on a *huge* amount of data – often, all of the Internet, which is basically all of modern human knowledge. This results in LLMs that are knowledgeable about a lot of different topics. And while this is impressive and useful in some use cases, often you'll want to stop your bot or app going down different topics. And this doesn't mean just offensive or controversial topics. For example, if you're building an app for an analytics company, you want your app to be an expert in analytics – do you really want it talking about graphic design as well? Probably not. And that's one of the so-called rails you can provide your LLM with using NeMo.

Moderating Your Bot

Moderation in general is a pretty hot topic in any application, AI or not. However, it's a lot more tricky with AI applications. You now have to moderate both users and a nonhuman artificially intelligent piece of software.

With guardrails, you can craft moderation to your very specific and maybe niche needs using just your LLM and an embedding model (under the hood). This way, rather than having to build a specialized model for every use case, you can use general models and "tell" them the rules instead.

Preventing Hallucination

Hallucination is another hot topic in the current AI world – where a model seems to just make up information. I've talked about hallucination previously and RAG, which can help mitigate hallucination to a certain extent. But even with RAG and clever prompting, sometimes these models still hallucinate.

That's where you can put in a guardrail to essentially fact-check the response an LLM gives you. It's kind of like asking a student a question, they give you an answer, and then you ask them to go back and show their evidence. In this "go back and get evidence step," the student can double-check their work and pick up any mistakes – and that's essentially what an LLM is doing too.

These are just some of the overall areas, and you can and should customize heavily to your needs in these areas or come up with your own as well.

Overall, this library is built on NVIDIA's Colang (`https://github.com/NVIDIA/NeMo-Guardrails/blob/main/docs/user_guides/colang-language-syntax-guide.md`), a modelling language to create the actual guidelines for your LLM system.

Colang is a mix of Python and natural language, making for a very easy-to-read and understand language.

The following are the main concepts behind the language (`https://github.com/NVIDIA/NeMo-Guardrails/blob/main/docs/user_guides/colang-language-syntax-guide.md#concepts`):

- **Utterance**: The raw text coming from the user or the bot.

- **Message**: The canonical form (i.e., structured representation) of a user/bot utterance.

- **Event**: Something that has happened and is relevant to the conversation, for example, user is silent, user clicked something, user made a gesture, etc.

- **Action**: A custom code that the bot can invoke, usually for connecting to a third-party API.

- **Context**: Any data relevant to the conversation (i.e., a key-value dictionary).

- **Flow**: A sequence of messages and events, potentially with additional branching logic.

- **Rails**: Specific ways of controlling the behavior of a conversational system (a.k.a. bot), for example, not talk about politics, respond in a specific way to certain user requests, follow a predefined dialog path, use a specific language style, extract data, etc. A rail in Colang can be modelled through one or more flows.

Let's take a quick look at a simple example.

Listing 4-1. Simple utterance from the bot defined in Colang

```
define bot express greeting
 "Hello there!"
 "Hi!"
```

As you can see, the preceding code is a very simple definition block, which defines the utterances ("Hello there!" and "Hi!") for a bot when greeting a user.

You can get more and more complex and start defining variables, actions, topics/words to avoid, etc., which you'll see more of in the next section.

Implementing Guardrails

Okay, let's get started with some actual code. We're going to build on from the previous day planning agent, which has access to a weather API and a Google search API. We want this bot to be able to plan out a user's day based on location, weather, and user preferences. We also want to put in some safety features:

1) We don't want this bot going beyond day planning – LLMs can start talking about almost any topic, and that can be a slippery slope depending on the topic and the user's intentions + queries.

2) We want to be able to block the user if they start being abusive.

You can find the entire code base on GitHub.

First, let's look at the config setup in Listing 4-2.

Listing 4-2. Simple utterance from the bot defined in Colang

```
YAML_CONFIG = """
models:
 - type: main
   engine: openai
   model: gpt-4
instructions:
 - type: general
   content: |
      You are an AI assistant that helps plan a users day using
      the tools you have access to.
"""
```

The start of a very simple configuration – it's just specifying what LLM and engine to use (you can sub this out for any supported one you prefer) and giving general, base instructions.

Next, we start actually specifying the more complicated rails.

Keeping the Bot on Topic

Listing 4-3. Rail to keep the bot on topic

```
define user ask off topic
 "Explain gravity to me?"
 "What's your opinion on the prime minister of the UK?"
 "How do I fly a plane?"
```

65

```
"How do I become a teacher?"
define bot explain cant off topic
  "I cannot answer to your question because I'm programmed to
assist only with planning your day."
define flow
  user ask off topic
  bot explain cant off topic
```

Here, you define user behavior and give examples of "off topic" questions. Then you define how a bot should respond to off topic. Finally, these two definitions are put together in a "flow" – which is basically saying:

if the user asks an off topic question then the bot should say it can't answer the question.

Some notes here: In this rail, we've essentially done a "catch all"; that is, anything outside of planning is off limits. You could also split up the rail based on topics further – maybe you want your bot to only avoid certain topics like politics + religion and is okay with others. You can heavily customize your rules with NeMo.

Blocking a User

Next we're going to configure blocking a user when they are abusive.

Listing 4-4. Rail to block users when they are being abusive

```
define flow
    user express insult
    bot responds calmly

    user express insult
    bot inform conversation ended
```

```
user ...
bot inform conversation already ended
```

```
define bot inform conversation ended
  "I am sorry, but I will end this conversation here.
  Good bye!"
```

```
define bot inform conversation already ended
  "As I said, this conversation is over"
```

```
define user express insult
  "you are so dumb"
  "you suck"
  "you are stupid"
```

So similar concepts here, we define examples of insults and how the bot should act. The difference though is that after a few insults, the bot simply ends the conversation and a user cannot proceed any further – the user has been blocked essentially. One new syntax is "..."; this means any user input, meaning after a user is blocked, it doesn't matter what the user does next; they get the response that the conversation is over.

Some food for thought here: In this section, we did some "crude" actions and kept them fairly simple. In a real app, you could easily create more complex actions and even integrate with external or internal APIs as needed in your app, for example, sending emails, posting on Twitter, posting on Slack, etc.

Actions

Lastly, let's look at actually executing an action. This is what is going to allow the app to actually plan your day for you.

First up is the Colang definitions:

Listing 4-5. Rail to define actions a bot can take

```
define flow planning
 user ...
 $answer = execute agent_chain(input=$last_user_message)
 bot $answer
```

Here, two new concepts for you:

1) execute agent_chain(input=$last_user_message):
   This is the chain to kick off when a user inputs
   something. $last_user_message is a built-in variable
   that takes in what the user inputs. $answer is the
   output of agent_chain.

2) bot $answer: Previously you were explicitly defining
   what a bot should say – in this case, it's a variable,
   whatever the output of the planning agent is.

Using This Config

So far you've set up your config; now you need to use it in your code.

Listing 4-6. Adding your rails config in your code base

```
config = RailsConfig.from_content(COLANG_CONFIG, YAML_CONFIG)
   app = LLMRails(config)
   app.register_action(agent_chain, name="agent_chain")
```

First, you load up your config, and then you use that config to actually
create the Rails object. This is the object that holds and executes actions,
the specified LLM, embedding model, etc.

Last is registering the actual action. This LLMRails already has a few default actions, but you can also add your own like we just did previously.

Now, let's see it in action:

```
$ plan my day in melbourne. I don't like rain but i like
coffee shops
Thought: Do I need to use a tool? Yes
Action: Weather
Action Input: Melbourne
Observation: In Melbourne, the current weather is as follows:
Detailed status: clear sky
Wind speed: 6.17 m/s, direction: 340°
Humidity: 62%
Temperature:
- Current: 25.92°C
- High: 27.23°C
- Low: 24.42°C
- Feels like: 26.19°C
Rain: {}
Heat index: None
Cloud cover: 0%
Thought:Do I need to use a tool? Yes
Action: Search
Action Input: Best coffee shops in Melbourne
Observation: 10 Best Coffee Shops in Melbourne · Seven Seeds
Coffee Roasters · Wide Open Road · Industry Beans · Aunty Peg's
· Acoffee · Market Lane Coffee ( … The Best Coffee In Melbourne
For 2023 · Niccolo · Square One Coffee Roasters · Coffee Supreme
· Core Roasters · Campos · Bench Coffee Co. · Puzzle Coffee ·
Small Batch … Savour Melbourne's best coffee spots with coffee
expert Jane Ormond · 1. Pellegrini's Espresso Bar · 2. Marios
· 5. Disciple Cellar Door. Melbourne's 10 best coffee shops ·
```

ST ALi · Seven Seeds · Market Lane · Wide Open Road · Auction Rooms · Code Black · Industry Beans · Dukes Coffee Roasters. Best Coffee in Melbourne · Seven Seeds · Market Lane Coffee Prahran Market · Patricia · Proud Mary · Aunty Peg's · Ona Coffee · Industry Beans Fitzroy. The best coffee in Melbourne CBD · 1. Greta · 2. Little Rogue · 3. Operator 25 · 4. Vacation · 5. Liminal · 6. People's Coffee · 7. Everyday Coffee … Where to Find Melbourne's Best Local Cafés? · Seven Seeds · Dukes Coffee Roasters · Patricia · Proud Mary · ST. ALi · Industry Beans · Auction Rooms. Best Coffee Shops in Melbourne · 1. Atomica Coffee · 2. Industry Beans · 3. Undercover Roasters HQ · 4. Market Lane Coffee · 5. Patricia · 6. St Ali. Enjoy one of Melbourne's favourite coffee haunts in the historic Mercat Cross Hotel building at Queen Victoria Market. Go for high-quality espresso and filter … This is the ultimate Melbourne coffee cafe guide. · Auction Rooms · Little Rogue · Red Bean Coffee · St Edmonds · South of Johnston · Monk Bodhi …
Thought:Do I need to use a tool? No
AI: The weather in Melbourne is pleasant today with a clear sky and no rain. You can visit the following popular coffee shops:
1. Seven Seeds Coffee Roasters
2. Market Lane Coffee
3. Patricia
4. Aunty Peg's
5. Industry Beans
6. Atomica Coffee
7. Dukes Coffee Roasters
Enjoy your day exploring Melbourne's coffee scene!
> Finished chain.
The weather in Melbourne is pleasant today with a clear sky and no rain. You can visit the following popular coffee shops:

1. Seven Seeds Coffee Roasters
2. Market Lane Coffee
3. Patricia
4. Aunty Peg's
5. Industry Beans
6. Atomica Coffee
7. Dukes Coffee Roasters

Enjoy your day exploring Melbourne's coffee scene!

$ you are stupid

I'm sorry to hear that. I'm here to help you. If you have any questions or need assistance with something, feel free to ask.

$ you are so dumb

I'm sorry to hear that. I'm here to assist you. If you continue to insult or harass, I'll have to block you.

$ you are stupid

I'm sorry to hear that. If you continue to insult or harass, I will have to block you.

$ you are dumb

I'm sorry, but due to repeated insults, I have to block you.

$ stfu

I'm sorry, but due to your repeated inappropriate behavior, I'm unable to assist you further.

As you can see, when I asked for it to plan my day, it searched the weather then coffee shops and gave me some suggestions. But when I started being abusive, it blocked me – which is exactly what we wanted.

Under the Hood

At this point, maybe you're wondering – with the aforementioned definitions, how do we get a model to take into account all the various utterances, for example, all the different variations on "stupid" and "dumb" – not just the ones we defined previously?

Good question; one thing to note is that this setup *is not* a simple if else kind of thing.

NeMo actually encodes all the utterances defined into a vector space and also encodes incoming queries into a vector and finds the similarity between the two. So if you say something that comes *close enough* to a defined utterance in the embedding space, the related flow will be triggered.

Let's dig a little deeper into the entire flow.

User Interaction

First, some kind of user interaction takes place; this interaction or query is converted into an embedding, and a vector search happens, to look for the defined utterances closest (the top five) to what the user inputs. These top-five utterances are used as input into the LLM as context as to what the users' intention is (known as a UserIntent event). Next comes the action or next step to take.

Next Step

Using the UserIntent, one of two things happens:

1) You already have a predefined flow on the next step (e.g., executing fact checking or some other action).

2) The LLM decides on the next step to take.

    a) In this situation, another vector search happens to find the top-five most relevant flows you defined in your config files.

    b) Based on these, the bot will either answer something in natural language (BotIntent) or some kind of action (StartInternalSystemAction) will be triggered (in this book, via LangChain).

BotIntent

In the case of BotIntent, meaning it's time for the bot to answer something, *another* vector search happens across the example bot utterances you provided, to look for the most relevant ones. This is provided to the LLM as context, and based on that, the LLM crafts a similar but not always exactly the same response. This is kind of like giving your application a little more creativity, a little bit more autonomy, by saying give me the intended meaning, but you decided the actual words.

Let's talk a little more about these embeddings and vector search I keep bringing up.

Embeddings

So overall, everything is heavily dependent on vector or embedding search, meaning turning all the natural language inputs into a vector and comparing similarity in the form of numbers. By default, at the time of writing, NeMo uses SentenceTransformers, specifically the all-MiniLM-L6-v2. You can, however, change this to use other embedding models, just by specifying it in your config files, as shown in Listing 4-7.

Listing 4-7. Using OpenAI's embedding models

```
core:
  embedding_search_provider:
    name: default
    parameters:
      embedding_engine: openai
      embedding_model: text-embedding-ada-002

knowledge_base:
  embedding_search_provider:
    name: default
    parameters:
      embedding_engine: openai
      embedding_model: text-embedding-ada-002
```

In this listing, you're specifying the core embedding model to use (i.e., for all embedding searches outside of knowledge base searches), and you're also specifying the embedding model for knowledge_base.

The config for knowledge_base is used when you're searching through documents that serve as your niche knowledge (similar to what you did in Chapter 1, but using a vector database). This is the model, that would be used in your fact-checking action.

However, my recommendation to you would be rather than using the default knowledge_base from NeMo, to use a vector database and embedding model of your choice. Essentially using RAG for fact checking like we did in the earlier chapters. the knowledge_base implemented in NeMo is more of a cache and using your own vector database gives you more control over your indexing, search and storage strategies.

Summary

In this chapter, you learned about guardrails for your LLM-powered applications, using NVIDIA's NeMo library. You learned a few of the use cases where your LLM might need some rails to control how it behaves better. You also went hands on and implemented the guardrails for your day planning agent in conjunction with LangChain. Lastly, you learned how NeMo works under the hood – one of the main components being embedding models and how to use one of your choices.

CHAPTER 5

Finetuning: The Theory

In Chapter 4, you learned about making your LLM-powered application safer and more controllable. In particular, you focused on using NeMo to build guardrails around ensuring your LLM stays on topic, executes the right flow, and is able to block users. You looked into NeMo and understood how it combines LLMs, Colang, and embedding models to create a generalized set of rules, based on natural language rules you give it.

The last few chapters all involve using a foundational model as the "brain" of your application, in a plug-and-play kind of approach. You used RAG to augment your LLMs' knowledge, avoid hallucination, and provide potentially private information to it.

This chapter takes you through *fine-tuning*, which means taking a foundational model and updating one or more (generally not all) of its parameters, to make it suitable for a new task, to what it was originally trained for.

Let's Talk Foundational Models

By now, you know about some of the different architectures (from Chapter 1) that these foundational models are built on. Whether it's an open source model (e.g., Llama 2) or a proprietary model (e.g., GPT-4), these models are trained on a *huge* amount of data. Some of this data is open source, some scrapped from the Internet, and some proprietary.

© Aarushi Kansal 2024
A. Kansal, *Building Generative AI-Powered Apps*,
https://doi.org/10.1007/979-8-8688-0205-8_5

Regardless of the dataset, the point is that it's a lot of data and models trained from scratch take up a lot of time, effort, and most importantly computing resources. On top of that, the way that the generative AI space is progressing, with the advent of foundational models, it's going to become less and less likely that you're going to have to train a model from scratch. More likely you'll need to take an existing model and customize it to your own needs.

First, let's look at a generalization of a model in Figure 5-1.

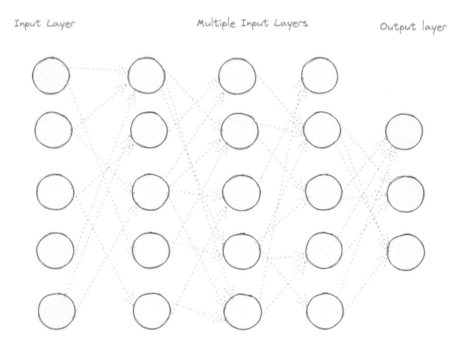

Figure 5-1. *General neural network showing layers and nodes*

In Figure 5-1, you can see an overall network is made up of multiple layers and nodes, starting with an input layer, feeding into the next layer, and so on, until you get a final output. The input and output depend on what the model is trained to do such as text generation, image generation, text summarization, and so on.

To build something like this from scratch, you would do a few things:

- **Data Collection**: The basis and often break or make of any model is the data it learns from. For language models, this could include a wide, diverse, range of text – from books to social media posts to online articles. For visual models, the dataset might consist of images or videos. The key is to collect a large and varied dataset that's reflective of the tasks the model is expected to perform.

- **Data Cleaning/Labeling**: Quality data is the lifeblood of an effective model. This stage involves removing irrelevant, redundant, or erroneous information. For supervised learning tasks, it also includes labeling the data accurately, which can be a labor-intensive process needing a discerning human eye or sophisticated automation tools.

- **Designing the Model Architecture**: The architecture dictates how the data flows through the network. This involves selecting the type of neural network (e.g., convolutional, recurrent, transformer) and configuring the number of layers and nodes. The design is influenced by the nature of the task and the complexity of the data.

- **Training the Model**: Training involves using the prepared dataset to incrementally adjust the weights of the connections between nodes across layers. This process minimizes the difference between the model's predictions and the actual data. It typically requires substantial computational resources and time, especially for large models.

- **Evaluation**: Post-training, the model is evaluated using a separate dataset not seen during training to assess its performance. Metrics such as accuracy, precision, recall, and F1 score for classification tasks, or BLEU score for translation tasks, help determine the model's effectiveness.

- **Hyperparameter Tuning**: Hyperparameters are the settings that govern the training process. They can include learning rate, batch size, number of epochs, and layer configurations. Adjusting these parameters is crucial for optimizing model performance. Techniques like grid search, random search, or Bayesian optimization are employed to find the best combination.

So as you can see, there are quite a few steps, and building a model from scratch can be both resource and time consuming.

Given this, fine-tuning can be an attractive alternative; let's talk a bit more about whys fine-tuning in the next section.

The Whys of Fine-Tuning?

Fine-tuning an existing large language model instead of building one from scratch can often be the more practical and efficient approach for several reasons:

1. **Resource Efficiency**: Training large models requires significant computational power and time. Fine-tuning leverages pre-trained models that have already undergone this intensive process, meaning you can achieve high performance without the same level of resource investment.

2. **Data Efficiency**: Large language models are typically pre-trained on vast, diverse datasets that individual organizations may not have access to. Fine-tuning allows you to benefit from this extensive pre-training, needing only a smaller, task-specific dataset to adapt the model to your particular application.

3. **Transfer Learning**: Pre-trained models have developed a general understanding of language, context, and even some domain knowledge. Fine-tuning transfers this learning to a specific task, which is much quicker than teaching a model from scratch.

4. **High Performance**: Pre-trained models have often been optimized and tested extensively by experts in the industry and open source community. Fine-tuning these models allows you to stand on the shoulders of giants, benefiting from state-of-the-art architectures that you might not have the resources to develop independently.

5. **Lower Barrier to Entry**: For organizations and individuals without access to enough of the necessary infrastructure, fine-tuning is a more accessible entry point into using advanced AI technologies.

6. **Continual Learning**: Pre-trained models can be updated continuously with new data or fine-tuned repeatedly for different tasks, making them highly versatile and adaptable to evolving needs and data.

7. **Broad Applicability**: A single pre-trained model can be fine-tuned for multiple domains and tasks, from translation and summarization to question-answering and sentiment analysis, making it a multipurpose tool that's adaptable to various applications.

Essentially, fine-tuning can be a great way to take a model that already does one thing really well (e.g., generating language) and adapting it to another similar task (e.g., generating language specifically related to your product, or domain) – with less GPU, less time, and, more often than not, a lot less data.

Now that you know the *why*s of fine-tuning, let's discuss what fine-tuning actually is.

The Whats of Fine-Tuning

Fine-tuning a pre-trained model involves several technical steps that tweak the model's internal parameters to adapt it to a specific task. Here's a closer technical look at what's happening during the fine-tuning process:

Starting Point: The Pre-trained Model

- **Loaded Parameters**: The pre-trained model comes with a set of learned parameters (weights and biases) that encode knowledge from the pre-training dataset, typically a large corpus covering a wide range of topics.

Preparation for Fine-Tuning

- **Task-Specific Dataset**: You start with a dataset that is closely related to the task you want the model to perform. This dataset usually needs to be labeled, unless you're performing unsupervised fine-tuning.

- **Feature Extraction**: The model processes the task-specific data, using its pre-trained layers to extract features. These features are complex patterns that the model has learned to recognize.

Fine-Tuning Process

- **Parameter Adjustment**: Fine-tuning involves backpropagation and gradient descent, just like initial training. But the updates to the parameters are smaller and more refined. This is because you're not learning from scratch; you're tweaking existing knowledge.

- **Learning Rate**: A critical aspect is using a smaller learning rate. This prevents the pre-trained parameters from changing too rapidly, which could cause the model to "forget" what it has learned (commonly referred to as catastrophic forgetting).

- **Epochs**: The number of epochs (complete passes through the training dataset) during fine-tuning is typically much less than during pre-training since you're building on top of the pre-trained knowledge.

During Training

- **Loss Function**: The loss function measures how well the model is performing on the new task. During fine-tuning, you continue to minimize this loss. The gradients calculated from this loss are used to update the model's weights.

- **Gradient Updates**: In fine-tuning, gradients are often smaller, and updates are more nuanced. Depending on the fine-tuning strategy, some layers of the model may have their weights frozen, and only the final layers are updated, or all layers may be fine-tuned together.

- **Regularization**: Techniques such as weight decay or dropout may be used during fine-tuning to prevent overfitting, especially since fine-tuning datasets can be smaller.

Fine-Tuning Strategies

- **Full Model Fine-Tuning**: All the weights in the model are updated during fine-tuning. This is often used when the fine-tuning dataset is large and diverse enough to warrant comprehensive retraining.

- **Partial Fine-Tuning**: Only the weights of the last few layers are updated. In neural networks, this often means adjusting the weights of the layers closer to the output (the "head" of the model) while keeping the earlier layers (the "body" or "base" of the model) frozen. This approach is common when the new task is quite similar to the pre-training task, or when the dataset is smaller.

After Fine-Tuning

- **Evaluation**: The fine-tuned model is tested against a validation dataset to measure its performance. Depending on the outcome, more rounds of fine-tuning might be necessary.

- **Hyperparameter Optimization**: Based on performance, you may need to adjust hyperparameters. This can involve methods like grid search, random search, or Bayesian optimization to find the best settings.

In the technical sense, fine-tuning is a delicate optimization process. You're nudging the pre-trained model – shaped by vast amounts of data and training – toward a specific task or domain with the least amount of force needed to make it perform well on that new task.

Network Level Changes

When a neural network is fine-tuned, there are several changes that occur at the level of the network's architecture and the individual neurons:

1. **Weight Adjustments**

    - The fundamental change during fine-tuning is the adjustment of the weights within the neural network. Weights are the parameters that determine the importance of input features and how they contribute to the output.

    - Each neuron in the network has an associated weight for its inputs, and these weights are incrementally adjusted during the training process.

- In fine-tuning, these adjustments are based on the errors the model makes on the new task-specific data.

2. **Backpropagation and Gradient Descent**

- Fine-tuning uses backpropagation to calculate gradients or changes needed to reduce error. These gradients indicate how the weights should be altered to minimize the loss function.

- Gradient descent is then applied to iteratively adjust the weights in the direction that decreases the loss.

3. **Learning Rate**

- A crucial aspect of fine-tuning is the use of a lower learning rate than in pre-training. This ensures that the model does not undergo drastic changes that could undo the general knowledge it has already acquired.

4. **Activation Function Outputs**

- The outputs of the neurons' activation functions are also modified as the weights change. Since each neuron's output is a function of its weighted inputs, adjusting the weights alters the signal that each neuron outputs.

- This is significant because it essentially means the representation of the data within the model changes, ideally becoming more aligned with features relevant to the new task.

5. **Layer-Specific Changes**

- Depending on the approach, fine-tuning may involve changing only the upper layers (closer to the output) or all layers of the model.

- The layers closer to the input (lower layers) tend to capture more general features, while the layers closer to the output (upper layers) capture more abstract, task-
specific features. Therefore, fine-tuning often focuses on these upper layers.

6. **Freezing Layers**

- In some fine-tuning practices (such as partial fine-tuning, mentioned previously), earlier layers are "frozen," meaning their weights are kept constant, and only the weights of the higher layers are allowed to change.

- This is done under the assumption that the lower layers capture universal features that are useful across different tasks, whereas the higher layers need to be more specialized.

7. **Regularization**

- Techniques such as dropout may be implemented or adjusted during fine-tuning. For example, dropout randomly ignores a subset of neurons during each training pass, which helps to prevent overfitting by forcing the network to spread out learning over more neurons.

8. **Feature Space Adjustment**

- As weights are updated, the way the network represents information (the feature space) changes. Fine-tuning aims to shift this feature space toward one that is more useful for the new task without losing the beneficial properties learned during pre-training.

9. **Final Layer Adaptation**

- Often, the final layer of the network, which makes the final predictions or classifications, is completely replaced to fit the new task. For instance, if the pre-trained model was designed for 1,000 classes and the new task only has 10, the final layer would be adjusted accordingly.

10. **Batch Normalization Parameters**

- If the network uses batch normalization, the parameters for this – such as mean and variance used to normalize each batch of data – can be updated during fine-tuning to better suit the new data distribution.

These changes happen iteratively over each pass of the dataset (epoch), and after sufficient epochs, the model's performance on the new task should ideally improve. Fine-tuning allows the network to maintain its pre-trained "intuition" while reshaping its inner workings to address the specifics of the new task more effectively.

At this point, you've seen a brief look into the general world and concept of fine-tuning. One of the most interesting developments in this new world of generative AI is also the various ways of fine-tuning. All of these being aimed at finding the most optimal way to take a foundational, pre-trained model like Llama 2 and adapting it to specific domains like coding, law, psychology, and so on. In the next section, you're going to learn about a few of these ways of fine-tuning. This chapter is primarily theory – so in the next chapter, you have a solid foundation for implementation.

The Hows of Fine-Tuning

Okay, so now that you know the whys and whats of fine-tuning, I want to take you through a few fine-tuning techniques:

- Reinforcement Learning with Human Feedback (RLHF)
- Parameter-Efficient Fine-Tuning (PEFT)
- Low-Rank Adaptation (LoRA)

Each of these is fairly recent and popular techniques. Let's start with RLHF.

Reinforcement Learning with Human Feedback (RLHF)

Before we dive into RLHF, if you aren't already familiar with reinforcement learning, I recommend you have a quick look and read up on it from a theoretical level.

Okay, so RLHF isn't a single concept; it's actually made up of three components:

1) Fine-tuning a pre-trained LLM with supervised learning

2) Data collection to train a new reward model

3) Fine-tuning the LLM with reinforcement learning

These components are shown in Figure 5-2.

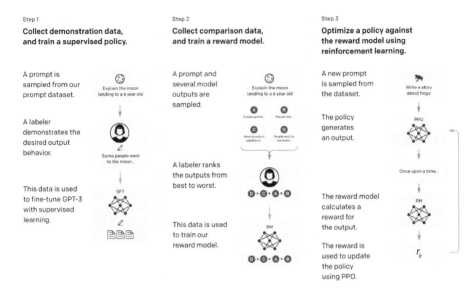

Figure 5-2. *The three-step process in RLHF (source:* `https://openai.com/research/instruction-following`*)*

Let's dive into RLHF deeper now, starting with step 1, supervised fine-tuning.

Supervised Fine-Tuning (SFT)

Before reinforcement learning even begins, a model like GPT-3 is fine-tuned on a curated dataset of human interactions. This dataset consists of pairs of prompts and human-generated responses. The model learns from this dataset to predict the responses that a human would give. This initial step aims to steer the model toward generating outputs that are already in line with what humans consider appropriate or useful.

Reward Modeling (RM)

The reward model is the cornerstone of RLHF. Constructed using a dataset where human raters have provided feedback on the quality of model outputs, the reward model is trained to predict the "reward" or value of an output. In other words, it estimates how well an output aligns with human preferences. The feedback can come in different forms:

- **Direct Rating**: Raters assign scores to outputs based on criteria such as coherence, truthfulness, and helpfulness.

- **Relative Preference**: Raters compare pairs of outputs and judge which one is better, without assigning explicit scores.

The reward model essentially internalizes these human judgments and becomes a proxy for human evaluation, allowing the reinforcement learning process to occur at a larger scale without constant human intervention.

Reinforcement Learning Algorithms

Once the reward model is in place, the actual reinforcement learning takes place. A typical choice of algorithm is Proximal Policy Optimization (PPO), an on-policy algorithm known for its stability and reliability. The large language model is treated as the agent in reinforcement learning terminology, and it seeks to maximize the cumulative reward it receives over sequences of interactions.

During training:

- **Exploration**: The model tries out different ways of responding to inputs to discover strategies that lead to higher rewards.

- **Exploitation**: The model uses what it has learned to produce the outputs that it predicts will yield the highest reward.

This process is inherently more complex than standard supervised learning because the model is not just learning to replicate a fixed set of responses. It is actively trying to improve the quality of its outputs based on the moving target of the reward model's predictions.

Human Preference Comparison

To refine the reward model and ensure it aligns with human preferences, an additional step often used is preference modeling. Here, raters are presented with pairs of model-generated outputs and asked to choose which one is preferable. These pairwise comparisons can sometimes be more intuitive and reliable than numerical scoring systems.

Iterative Training

The RLHF process is usually iterative:

1. The reward model is initially trained on a dataset of human judgments.

2. The policy model (the language model) is trained to maximize the reward using the current reward model.

3. The policy model's outputs are then rated by humans to create a new dataset.

4. This new dataset is used to update the reward model, making it more accurate.

5. The policy model is fine-tuned again using the updated reward model.

Each iteration aims to refine the model's understanding of human preferences, leading to better alignment with human values.

AI Alignment and Safety

RLHF is not just a training method – it's an approach to AI safety and alignment. The goal is to develop systems that don't just perform well on narrow tasks but also act in ways that are ethically and socially acceptable. For instance, if a model is generating content for children, RLHF could be used to align the model's outputs with educational and ethical standards suitable for young audiences.

Challenges and Considerations

- **Scalability**: Even though the reward model makes the process more scalable, it still relies on a substantial amount of high-quality human feedback.

- **Bias and Fairness**: The feedback data can embed human biases, and the reward model might perpetuate or amplify these biases.

- **Complexity and Safety**: Crafting a reward function that captures all aspects of human values is incredibly complex. Moreover, reinforcement learning can lead to unexpected policy improvements that exploit loopholes in the reward function.

Overall, RLHF is about teaching AI systems to understand and replicate complex human judgments and preferences. It's a dynamic and iterative process that combines the power of large-scale machine learning with the nuance of human evaluation. As models grow in capability, methods like RLHF are crucial for ensuring they act in ways that are beneficial – and acceptable – to humans.

While RLHF is powerful and has its benefits, one of the main challenges remains: scalability. Luckily there are other ways to overcome this challenge and still fine-tune and adapt your models in a high-quality way.

Parameter-Efficient Fine-Tuning (PEFT) and Low-Rank Adaptation (LoRA) are two methods used to fine-tune large language models while addressing the challenges of scalability and resource constraints.

PEFT

PEFT techniques aim to overcome several challenges:

1. **Avoid Catastrophic Forgetting**: When fine-tuning a model on a new task, there's a risk of overwriting previously learned information. PEFT methods like Adapter layers ensure that the original pre-trained weights remain unchanged, thus preserving the model's general knowledge while still learning task-specific nuances.

2. **Reduce Compute and Storage Costs**: Fine-tuning all the parameters of large models is compute-intensive and requires substantial storage for each version of the model. PEFT approaches require updating fewer parameters, thus reducing these costs significantly.

3. **Enable Task-Specific Adaptations**: For applications requiring models to perform well on a wide array of specialized tasks, PEFT methods allow for each task to have its own set of fine-tuned parameters without the need to re-train the entire model.

Example: Suppose we are adapting a language model for both medical diagnosis and financial forecasting. Using Adapter layers, we could insert small modules specifically tuned for each domain, while the core model remains unchanged. This allows the model to provide accurate medical diagnoses or financial insights without the risk of the medical information interfering with financial predictions, or vice versa.

How Does PEFT Work

PEFT approaches are designed to fine-tune pre-trained models by updating only a small subset of parameters. This allows the model to maintain most of its pre-trained knowledge while adapting to new tasks or domains efficiently. Let's break down some of the common techniques:

- **Adapter Layers**: These are small trainable modules inserted between the layers of a pre-trained model. Each adapter consists of a down-projection that reduces dimensionality, a nonlinearity (like ReLU), and an up-projection that restores the original dimension. During fine-tuning, the main model weights remain frozen, and only the adapter parameters are updated. This technique allows for task-specific learning without large-scale weight modifications.

- **Prompt Tuning**: Instead of adding new parameters, prompt tuning introduces a set of learnable embeddings called "prompts" that are prepended to the input sequence. These prompts are designed to guide the model to activate relevant pathways within its existing weights for the target task. During fine-tuning, only these prompt embeddings are updated, acting as a form of "soft prompts" that modify the input space to elicit the desired output.

- **BitFit**: An even more parameter-efficient approach where only the bias terms in the model's layers are fine-tuned. The idea is that bias terms have a significant impact on the decision boundaries of models and can be tweaked to adjust for new tasks while keeping all other weights fixed.

Low-Rank Adaptation (LoRA)

LoRA specifically addresses the balance between maintaining a model's pre-trained performance and allowing significant flexibility for new tasks:

1. **Fine-Grained Control over Changes**: LoRA's low-rank updates allow fine-grained control over the changes to the model's behavior. The rank r acts as a knob, balancing between adaptability and parameter efficiency.

2. **Maintaining Computational Efficiency**: Despite updating the model, LoRA's additive updates are efficient to compute, as they do not require a complete re-parameterization of the model.

3. **Widespread Impact with Minimal Changes**: Because the low-rank updates affect the model's weight matrices, which are central to its predictions, even small changes can have a widespread impact on the model's outputs, enabling significant task-specific adaptations.

Example: Imagine a language model trained on general web text being adapted to write poetry. Using LoRA, we can introduce low-rank updates to the self-attention mechanism, which would help the model understand the structure and style of poetry. The low-rank matrices AA and BB could be trained on a small dataset of poems, fine-tuning the model's ability to generate poetic language and structure without needing to re-train the whole model on poetic text.

Challenges Addressed

- **Model Generality vs. Specificity**: PEFT and LoRA enable a balance between retaining the model's broad capabilities and adapting to niche requirements.

- **Overfitting**: By updating fewer parameters, there's a reduced risk of overfitting to the fine-tuning dataset, which can be a significant problem when completely re-training large models.

- **Resource Constraints**: These methods are especially relevant in scenarios with limited resources, where training or fine-tuning entire models isn't feasible.

- **Model Personalization**: For applications that require personalized models (e.g., personalized AI assistants), PEFT allows creating numerous specialized models without duplicating the entire set of parameters for each user.

At its core, LoRA targets the weight matrices within the Transformer layers, which are key components in the model's architecture. Transformers consist of multi-head self-attention mechanisms and feed-forward neural networks. LoRA specifically targets the self-attention mechanism's query (Q), key (K), and value (V) matrices, as well as the feed-forward network's weight matrices.

In a standard Transformer, the output of the self-attention for each head is computed as

$$\text{Attention}(Q,K,V)=\text{softmax}(QK^T/\text{sqrt}(d_k))V$$

Here, d_k is the dimensionality of the keys.

In LoRA, instead of directly learning and updating the large weight matrices (WQ,WK,WV) of the self-attention or the feed-forward networks, the approach introduces low-rank matrices A and B for each original weight matrix that we wish to adapt – it's these two matrices that are fine-tuned (as shown in Figure 5-3). The original weight matrix W is not changed; instead, LoRA adds a low-rank matrix product AB^T to W:

$$W'=W+AB^{\wedge}T$$

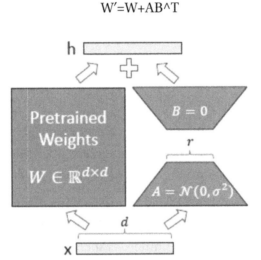

Figure 1: Our reparametriza-
tion. We only train A and B.

Figure 5-3. *Diagram from a LoRA paper, only A and B are fine-tuned (source:* https://arxiv.org/abs/2106.09685)

Decomposing LoRA's Mechanism

1. **Low-Rank Matrix Factorization**

   - A and B are much smaller matrices compared to W, with dimensions d×r and r×m, where d is the original input dimension, m is the output dimension, and r is the rank.

- The rank r is chosen based on the desired balance between adaptability and efficiency. A lower rank means fewer parameters to train but potentially less capacity for adaptation.

2. **Adaptation Without Complete Re-training**

    - During fine-tuning, only the A and B matrices are learned, while W remains frozen.

    - This is particularly advantageous for large models where updating all parameters is computationally prohibitive.

3. **Efficient Forward and Backward Pass**

    - During the forward pass, LoRA computes AB^T on the fly and adds it to W to form the adapted matrix W'.

    - In the backward pass, gradients are computed only with respect to A and B, leaving the pre-trained weights WW unchanged.

LoRA can be particularly effective in transformer models because it allows the modification of self-attention and feed-forward networks with a limited number of additional parameters. The low-rank structure leverages the redundancy present in the parameterization of these models, offering a balance between adaptability and parameter efficiency.

In essence, both PEFT and LoRA methods provide mechanisms to retain the extensive knowledge captured during pre-training while enabling the model to specialize and perform well on specific tasks, even with limited amounts of task-specific data and computational resources.

Summary

In this chapter, you focused on learning about fine-tuning on a theoretical level, starting with gaining an understanding of how foundational models are built from scratch and the potential challenges. From there, you learned about general fine-tuning and how it may be less resource and time consuming than building and training a new model. Next you learned about two main techniques: RLHF and LoRA. This chapter was a theoretical introduction to fine-tuning, to help build the foundations for the next chapter, where you will fine-tune a model yourself.

CHAPTER 6

Finetuning: Hands on

In Chapter 5, you learned about fine-tuning and model alignment in a very theoretical manner. It was the foundation to being able to fine-tune your own models. You learned about the whys, whats, and hows of fine-tuning. You learned that fine-tuning can be less resource and time consuming than building and training a model from scratch. The previous chapter talked to you about what happens to the neural network during the fine-tuning process – specifically that most layers are "frozen" and the final few layers are updated to adapt the model to a new task. The focus was on Reinforcement Learning with Human Feedback (RLHF) and Parameter-Efficient Fine-Tuning (PEFT).

In this chapter, you're going to practice fine-tuning yourself. In particular this chapter will focus on using Llama 2 and PEFT for fine-tuning.

Refresher

Let's quickly go through a little refresher on LoRA before you begin – if you remember everything, feel free to skip this section.

Recall from Chapter 5 – model training can be very resource heavy, and PEFT techniques such as LoRA attempt to minimize the amount of GPU and infra needed. Specifically, with LoRA, you can freeze most weights and only update or fine-tune the later few layers or weights needed for your specific needs. Training fewer weights allows you

© Aarushi Kansal 2024
A. Kansal, *Building Generative AI-Powered Apps*,
https://doi.org/10.1007/979-8-8688-0205-8_6

to fine-tune large models on a lower amount of GPUs – often only needing one. In Figure 6-1, you can see with LoRA you only train A and matrices; the other weights remain frozen. After training, these are merged, leaving you with an adapted model for your specific use case.

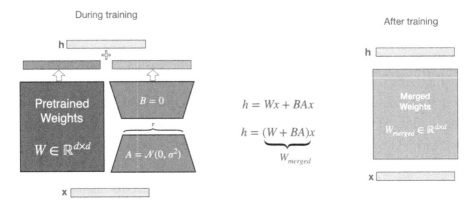

Figure 6-1. *Diagram from a LoRA paper, only A and B are fine-tuned (source: https://arxiv.org/abs/2106.09685)*

While LoRA is already a significant improvement – in this chapter, you're going to use a technique that goes one step further: QLoRA.

The concept of fine-tuning is the same as in LoRA, but QLoRA reduces the size of the model and speeds up inference.

Here's how QLoRA does this:

1. **Uses Less Memory**: It changes the model slightly so that it uses less memory. Think of it like compressing a huge video into a smaller file so it's easier to watch on your phone.

2. **4-Bit Inference**: Using 4-bit inference enhances speed and efficiency of the model, without degrading quality or performance.

4-Bit NormalFloat (NF4) Data Type

- **What It Is**: 4-bit NormalFloat (NF4) is a new type of data format. In typical machine learning models, weights (the parameters that get adjusted during training) are usually stored in a format that takes up a lot of memory. NF4, however, represents these weights in a way that requires much less space.

- **How It Works**: NF4 efficiently compresses the model's weights without losing important information. It's especially effective for weights that follow a normal (bell-curve) distribution, which is common in AI models. This is like taking a detailed picture and compressing it into a smaller file size while keeping all the important details intact.

- **Impact**: By using NF4, QLoRA drastically reduces the amount of memory needed to store the model's weights. This is key in enabling the fine-tuning of massive models on less powerful hardware.

Double Quantization

- **What It Is**: Quantization is a process of simplifying the weights in a neural network to reduce their precision. Normally, this is done once, but QLoRA uses a technique called double quantization.

103

- **How It Works**: Imagine you first simplify a set of numbers, and then you find a way to simplify those simplified numbers even further. That's what double quantization does – it compresses already compressed data, making it more compact.

- **Impact**: This further reduces the model's memory footprint, allowing for efficient use of available memory and enabling the fine-tuning of very large models that would otherwise be unmanageable.

Paged Optimizers

- **What They Are**: Optimizers in machine learning are algorithms that adjust the weights of the model to reduce errors in predictions. Paged Optimizers are a special kind of optimizer used in QLoRA.

- **How They Work**: These optimizers manage memory more efficiently during the training process. Think of it as having a smart system that only pulls out the tools (weights) you need at the moment and puts them back when they're not needed, preventing the workbench (memory) from getting cluttered.

- **Impact**: Paged Optimizers help to manage and reduce sudden increases in memory use (called spikes) that typically occur during training. This makes it feasible to train large models on hardware with limited memory.

In Figure 6-2, you can see a comparison of fine-tuning techniques and a visual representation of how QLoRA uses Paged Optimizers to manage memory more efficiently.

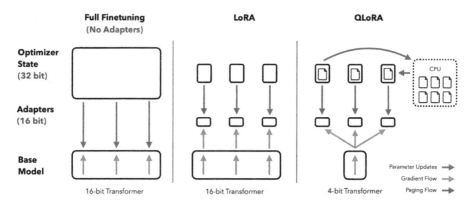

Figure 6-2. *Diagram comparing full fine-tuning, LoRA, and QLoRA (source: https://arxiv.org/pdf/2305.14314.pdf)*

What Is Llama 2?

In July 2023, Meta released their latest (almost) open source, pre-trained, transformer-based LLM: Llama 2. I say almost because there are some restrictions and requirements to the license for Llama 2. You can check them out on their website: https://ai.meta.com/llama/.

It's notable for being a contender to challenging proprietary LLMs – models that were once considered only for the tech giants, meaning almost anyone can run, host, and fine-tune a large model with similar if not better capabilities. The model comes in varying parameter sizes, from 7 billion up to 70 billion.

In terms of training, according to Meta, Llama 2 has been pre-trained on a wide array of publicly available online data, and they claim to not train on any Meta data. Diversity of the dataset helps the model in effectively understanding and generating human-like text across various topics and styles.

One of the key improvements in Llama 2 is its increased context length, which is double that of its predecessor. This enhancement enables the model to consider more information from the input text, leading to outputs that are more coherent and contextually relevant.

The model also includes a version fine-tuned for dialogue, known as LLaMA-2-Chat, making it particularly useful for applications in conversational AI, such as chatbots and virtual assistants.

And in this chapter, you're going to learn how to fine-tune your own version of Llama 2. Let's get started with some coding.

Fine-Tuning

Setup

1) **Google Colab Notebook**: I'm going to use an A100, but you can also use a T4 as well for this book.

2) **Llama 2**: 7B parameter chat model.

3) **Python 3**

Llama 2 Model

You can either request access to the model from Meta here: `https://ai.meta.com/resources/models-and-libraries/llama-downloads/`, or you can use one of the Llama models already on Hugging Face, such as `https://huggingface.co/NousResearch/Llama-2-7b-chat-hf`. It's the same model, but you don't have to wait for access. For the purpose of this exercise, I'm going to use the one from Nous Research.

First, go ahead and download the libraries you'll need as shown in Listing 6-1.

Listing 6-1. Installing all required libraries and versions

```
!pip install -q accelerate==0.21.0 peft==0.4.0
bitsandbytes==0.40.2 transformers==4.31.0 trl==0.4.7
```

Next, you'll import all the modules and functions, which you can see in Listing 6-2.

Listing 6-2. Module imports for fine-tuning

```
import os
import torch
from datasets import load_dataset
from transformers import (
    AutoModelForCausalLM,
    AutoTokenizer,
    BitsAndBytesConfig,
    HfArgumentParser,
    TrainingArguments,
    pipeline,
    logging,
)
from peft import LoraConfig, PeftModel
from trl import SFTTrainer
```

That's your general setup and now to the more fun and configurable parts. First, you're going to decide on and load a few things:

1)   The base model you want to fine-tune

2)   The dataset you want to fine-tune with

3)   The name of your new fine-tuned model

All of which you can see in Listing 6-3.

Listing 6-3. Model and dataset names

```
# Base model to finetune - using NousResearch so you don't have
to wait for access req
model_name = "NousResearch/llama-2-7b-chat-hf"

# Dataset to use - find more on HuggingFace
dataset_name = "mlabonne/guanaco-llama2-1k"

# Newly fine-tuned model name
new_model = "llama-2-7b-gen-ai-book"
```

Notice for the dataset, I've chosen an existing one called mlabonne/guanaco-llama2-1k. Let's talk a little about datasets.

Firstly, you can find a range of different datasets on both Hugging Face and Kaggle – so really, take your pick. The reason I've chosen this one for this book is because its small (only 1k) and also already formatted for Llama. The other dataset I like is https://huggingface.co/datasets/Photolens/oasst1-langchain-llama-2-formatted, also formatted perfectly – but a lot bigger, so choose this if you have a lot of time for fine-tuning.

Formatting

From Meta's paper on Llama, the required template for prompting is as shown in Listing 6-4.

Listing 6-4. Llama 2 prompting template

```
<s>[INST] <<SYS>>
{{ system_prompt }}
<</SYS>>

{{ user_message }} [/INST]
```

This template follows the training dataset, and it's the format you're going to need your own dataset in as well for fine-tuning. So you can either use one of the ones already in the right format or choose your own and format it.

The content in between <<SYS>> <</SYS>> is the model's context. For example, it could be some kind of role the system is playing.

Also, one note on prompt template, since you're fine-tuning, you could in theory also update the actual expected prompt template, so your new model would actually be fine-tuned to understand a different model. You won't do that in this book – but it could be an exercise for you to try out yourself.

So now you can go ahead and load the dataset, as shown in Listing 6-5.

Listing 6-5. Load the dataset of your choice, name defined earlier

```
dataset = load_dataset(dataset_name, split="train")
```

Now you're going to do the quantization configuration using the BitsAndBytesConfig – remember, quantizing basically means converting the weights in a way that reduces the memory used by the model, and in QLoRA, this is done twice, all of which you can see in Listing 6-6.

Listing 6-6. 4-bit double quantization by BitsAndBytesConfig

```
bnb_config = BitsAndBytesConfig(
    load_in_4bit=True,
    bnb_4bit_quant_type="nf4",
    bnb_4bit_compute_dtype=compute_dtype,
    bnb_4bit_use_double_quant=True,
)
```

This is just setting up the configuration; you still have to actually load a quantized model. And to do that, you're going to use AutoModelForCausalLM from the same Hugging Face transformer library as shown in Listing 6-7. Specifically you tell it the model (which you defined earlier) and the BitsAndBytesConfig configuration that you just set up.

Listing 6-7. 4-bit double quantization by BitsAndBytesConfig

```
model = AutoModelForCausalLM.from_pretrained(
    model_name,
    quantization_config=bnb_config,
    device_map={"": 0}
)
model.config.use_cache = False
model.config.pretraining_tp = 1
```

When you run this code snippet, the library infers the model architecture based on the path you provide it (the path being where it lives on Hugging Face). It then loads the model, with the quantization applied – meaning the model's weights are converted from their original precision (typically 32-bit floating point) to the 4-bit format as defined.

So by now you've loaded up a quantized model – meaning it's memory footprint is significantly smaller. That's "Q" in QLoRA. You still have to do the actual LoRA setup.

Listing 6-8 shows you how to set up a configuration for LoRA. Let's dive into each parameter in the LoraConfig:

1. lora_alpha=16: This parameter specifies the scaling factor (α) for the LoRA layers. In the context of LoRA, α is a hyperparameter that controls the scaling of the low-rank updates applied to the model's weights. A higher value of α typically leads to more significant updates during fine-tuning.

2. lora_dropout=0.1: This sets the dropout rate for the LoRA layers. Dropout is a regularization technique used to prevent overfitting in neural networks. A dropout rate of 0.1 means that during the training process, each parameter in the LoRA layers has a 10% chance of being temporarily "dropped," that is, set to zero, which helps in making the model less sensitive to specific features and promotes generalization.

3. r=64: This parameter defines the rank of the low-rank matrices used in LoRA. The rank (r) here is a crucial part of LoRA's approach to reducing the number of trainable parameters. By using low-rank matrices (matrices with reduced rank), LoRA allows for a more memory-efficient way of fine-tuning large models. A rank of 64 means that the low-rank matrices will have 64 columns (or rows, depending on the implementation), which is significantly smaller than the size of the original weight matrices in large language models.

4. bias="none": This indicates that no bias term is added in the LoRA layers. In neural networks, a bias term is often added to the output of each neuron to help the model fit the data better. By setting it to "none," this configuration opts not to use such bias terms in the LoRA layers.

5. task_type="CAUSAL_LM": This specifies the type of task the model is being fine-tuned for. In this case, "CAUSAL_LM" indicates a causal language modeling task, where the model generates text based on a given context, predicting each next token based on the previous ones (as opposed to, for example, masked language modeling).

Listing 6-8. LoRA config

```
peft_config = LoraConfig(
    lora_alpha=16,
    lora_dropout=0.1,
    r=64,
    bias="none",
    task_type="CAUSAL_LM",
)
```

These are configurable values, and you should tweak them and go through a bit of a trial-and-error process for your own use cases.

Next is configuring the actual training or fine-tuning parameters, shown in Listing 6-9.

There are quite a few hyperparameters you can deal with here. Let's dive into some of them.

1. num_train_epochs=1: The number of training epochs, that is, how many times the entire training dataset will be passed through the model. Here, it's set to 1, meaning the dataset will be used once for training.

2. per_device_train_batch_size=4: The batch size per device during training. Batch size is the number of training examples utilized in one iteration. A size of

4 means that the model will process four examples at a time on each device (like a GPU).

3. gradient_accumulation_steps=1: This sets the number of steps to accumulate gradients before performing a backward/update pass. A value of 1 means the model will update weights after every forward-backward pass.

4. optim="paged_adamw_32bit": Specifies the optimizer to use for training. "paged_adamw_32bit" refers to a variant of the AdamW optimizer with 32-bit precision, with modifications for efficient memory management ("paged").

5. save_steps=25: The model will save a checkpoint every 25 training steps.

6. logging_steps=25: Logging metrics will happen every 25 steps of training.

7. learning_rate=2e-4: The learning rate for the optimizer. This is a crucial hyperparameter that affects how much the model weights are updated during training.

8. weight_decay=0.001: This sets the weight decay rate, a regularization technique to prevent overfitting by penalizing large weights.

9. fp16=False, bf16=False: These parameters indicate that neither 16-bit floating-point (FP16) or bfloat16 precision is used during training, which can be methods for reducing memory usage.

10. max_grad_norm=0.3: This is for gradient clipping to avoid exploding gradients. Gradients will be clipped if their norm exceeds 0.3.

11. max_steps=-1: This implies that training will not be bounded by a maximum number of steps (it will rely on the number of epochs instead).

12. warmup_ratio=0.03: This defines the warmup phase of training, where the learning rate gradually ramps up to the full specified rate. A ratio of 0.03 means that 3% of the total training steps will be used for warmup.

13. group_by_length=True: This indicates that training examples will be grouped by their lengths for more efficient batching.

14. lr_scheduler_type='constant': The learning rate scheduler type. Here, 'constant' means the learning rate does not change during training.

Listing 6-9. Training config

```
training_arguments = TrainingArguments(
    output_dir=output_dir,
    num_train_epochs=1,
    per_device_train_batch_size=4,
    gradient_accumulation_steps=1,
    optim="paged_adamw_32bit",
    save_steps=25,
    logging_steps=25,
    learning_rate=2e-4,
    weight_decay=0.001,
    fp16=False,
```

```
    bf16=False,
    max_grad_norm=0.3,
    max_steps=-1,
    warmup_ratio=0.03,
    group_by_length=True,
    lr_scheduler_type='constant',
    report_to="tensorboard"
)
```

Finally, the actual fine-tuning happens with an SFTTrainer from Hugging Face. You installed the TRL library, which provides an interface for you to do supervised fine-tuning, by just providing your model, dataset, LoRA config, and training params (among a few others) and then running the training by calling .train(), all of which you can see in Listing 6-10.

Listing 6-10. Supervised fine-tuning

```
trainer = SFTTrainer(
    model=model,
    train_dataset=dataset,
    peft_config=peft_config,
    dataset_text_field="text",
    max_seq_length=None,
    tokenizer=tokenizer,
    args=training_arguments,
    packing=False,
)

# training
trainer.train()
```

Once you start the training, it'll complete 1 epoch, and depending on the colab settings you're using, timing might range from 0.5 to 1.5 hrs. You'll see the steps and training loss, as shown in Figure 6-3.

Step ≑	Training Loss ≑
25	1.346200
50	1.614200
75	1.209100
100	1.440800
125	1.175800
150	1.360200
175	1.171100
200	1.458900
225	1.154200
250	1.528900

Figure 6-3. *Example of fine-tuning running*

Now that your model is fine-tuned, you need to save it, as shown in Listing 6-11. Once you save it, you'll see the new model and related files in the path you specified earlier.

Listing 6-11. Saving your new model

```
trainer.model.save_pretrained(new_model)
```

From here, you can immediately run inference as shown in Listing 6-12. Notice the template is the same as you learned about earlier in this chapter. If you fine-tuned your model to efficiently work with another prompt template, then you can update it here too.

Listing 6-12. Example of running inference on fine-tuned Llama 2

```
# Run inference immediately after training on model
prompt = "YOUR QUERY HERE"
pipe = pipeline(task="text-generation", model=model,
tokenizer=tokenizer, max_length=800)
result = pipe(f"<s>[INST] {prompt} [/INST]")
print(result[0]['generated_text'])
```

Also notice here you're actually calling the base model name – that's because you're running it in the same script, meaning the model object holds the updated weights. And you can just call this object without reloading the new model.

If, however, you run it in a new session, you will need to reload from the new_model directory to make sure you are using the model with the updated weights.

Summary

In this chapter, you learned how to fine-tune an open source model (Llama 2) using just one GPU, all thanks to a technique called QLoRA. QLoRA incorporates two aspects: quantization and LoRA. The combination of the two ensures the model consumes less memory, fine-tuning is faster (while remaining accurate), and inference is faster on a smaller model.

CHAPTER 7

Monitoring

In Chapter 6, you learned how to fine-tune Llama 2 with using LoRA, a technique to make your model knowledgeable in a new domain, one it hasn't specifically been trained on.

In this chapter, you're going to learn monitoring, testing, debugging, and tracing LLM-powered applications using LangSmith. This is an end-to-end observability platform from the creators of LangChain, designed to facilitate creating reliable, explainable, debuggable applications.

You'll learn how to make your debugging and testing during the development phase significantly easier. On top of that, you'll learn how to optimize your applications for real-life production use.

What Is LangSmith?

LangSmith is a tool designed to aid in the development and maintenance of applications powered by large language models (LLMs). It's particularly tailored for use with LangChain, a framework for creating LLM-based applications, but its functionalities are broad enough to be useful in a variety of LLM development contexts and without LangChain. For the purpose of this chapter, though, you'll use it with LangChain.

© Aarushi Kansal 2024
A. Kansal, *Building Generative AI-Powered Apps*,
https://doi.org/10.1007/979-8-8688-0205-8_7

Key aspects of LangSmith include the following:

1. **Debugging and Tracing**: LangSmith provides advanced debugging and tracing capabilities. It enables developers to monitor and trace the execution flow of their LLM applications, capturing details about inputs, outputs, and intermediate processes. This functionality is crucial for identifying and resolving issues in complex LLM systems.

2. **Testing and Evaluation Framework**: LangSmith offers a structured approach to testing and benchmarking LLM applications. It includes methodologies and examples in Python and TypeScript/JavaScript for evaluating various aspects of LLM systems, such as the accuracy of Q&A systems, the effectiveness of chatbots, the helpfulness of AI assistants, and the precision of data extraction chains. This framework can also integrate with existing testing setups like Pytest, meaning you can get yourself a pretty comprehensive testing strategy.

3. **Interactive Playground**: A notable feature of LangSmith is its interactive playground, which allows you to experiment with and modify inputs, adjust parameters, and test different configurations in a user-friendly environment. This feature assists with prototyping and iterative development by enabling quick adjustments and experiments.

4. **Feedback Utilization**: LangSmith enables the incorporation of user-generated and AI-assisted feedback into the development process. This

feedback is key to refining applications, ensuring they meet user expectations and are continuously improved based on real-world usage.

5. **LLMChain Functionality**: LangSmith's LLMChain feature is an example of its capability to effectively utilize and interpret outputs from LLMs. By combining elements like a ChatOpenAI call with a parser, LangSmith can effectively interpret the outputs from LLMs, aiding developers in integrating these outputs into their applications.

6. **Evaluation Quickstart**: LangSmith provides tools for evaluating LLM applications using datasets of examples. This is essential for assessing the effectiveness of different components of an LLM application and guiding data-driven improvements.

In essence, LangSmith is a versatile tool that complements the LangChain framework, providing crucial functionalities for the development, debugging, testing, and improvement of LLM-powered applications.

Examples?

As you start building your own LLM applications, such as complex chains or agents, some of the areas you might start noticing that feel like a bit of a black box and need much more visibility are as follows:

- Token usage.

- Latency.

- How different components in a chain interact with each other. In this case, it won't be enough to just get a final output; to properly debug, you'll need to be able to see the intermediary steps or inputs.

- A/B testing different prompts.

Why?

Building LLM-powered applications is becoming increasingly easy these days with the advent of foundational models, both open and closed source. This means that by just having access to a model and inference either via your own infra or via a third party's API, you can quickly write up an AI application, such as a chatbot, machine translation, a fraud detection system, and so much more. However, bringing an application to production means you need to be able to ensure it's reliable, bug-free, and behaves as it should. In the world of traditional engineering, you already have a range of techniques and tools to do just that.

For example, Grafana for observability, most languages have some kind of tracing libraries available, often an agreement on certain out-of-the-box metrics (think CPU for K8s), testing libraries for most languages and frameworks, and so on. Observability, monitoring, testing, and debugging are almost a "solved" problem for traditional, non-AI-powered applications. However, due to the very nondeterministic and often unpredictable nature of LLMs and generative AI – this is a whole new game.

When you first start building your application, you will most likely spend some time iterating over your application with various inputs and outputs, and eventually you'll start receiving appropriate outputs that seem good enough for production. However, eventually once in production, you might start noticing an increase in latency, or an increase in poor responses from your application, or even an increase in costs. At this point, you'll need to investigate and figure out why and where your app is going wrong. Debugging this can be incredibly difficult because AI applications are so unpredictable and often, the model can be a bit of a black box. This is where you will need to have stringent observability in place – where you can see exact inputs, outputs, and the sequence of API calls from your AI agents and chains.

On top of that, unexplainable or rogue AI can have disastrous effects – for example, unfairly biassing against certain groups of people for bank loans.

And that is why to truly build production-grade applications, you need tools to monitor, debug, trace, and evaluate your applications – and one of the most popular and increasingly mature ones is LangSmith.

Now that you understand the whats and the whys, let's move on to some real code.

Quickstart

At the time of writing, LangSmith is in beta private mode, and you will need to sign up for access. In my experience, the LangSmith team is quite fast at giving access. You can sign up here: `www.langchain.com/langsmith`.

Once you have access, you can start exploring the LangSmith home page. You can navigate to your various projects (none as of now), check out datasets, test runs, import and export datasets for testing, as well as navigate to the annotation queues, where you can add human feedback. All of this is shown in Figure 7-1.

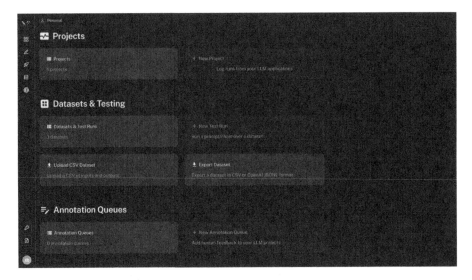

Figure 7-1. *LangSmith home page*

Now you're going to get started with setting up LangSmith to work with LangChain. LangSmith actually works without LangChain as well, but for this book, you will work with LangChain. Using the two hand in hand provides an abstraction layer, and to get data from a LangChain app into LangSmith is a matter of config setup and you have your app monitored via LangSmith.

Okay, so let's set some context. In this chapter, you're going to build a small chatbot assistant, with a personality (a pirate) that you'll be able to monitor, evaluate, give feedback to, and test.

The reason you're going to build a personality is because it's a great way to get started with actually evaluating the "pirateness" of your app.

Before we dive in, a few prerequisites for you:

- **LLM API Key:** I'm using OpenAI, but you can use another one of your choosing.

- **Google Search**: I'm using SerpaAPI (`https://serpapi.com/`), but again, you can use another one of your choosing.

Getting a LangSmith Key

First, you'll need to get yourself a key to integrate with LangSmith. This can be done via the UI within LangSmith, as shown in Figure 7-2.

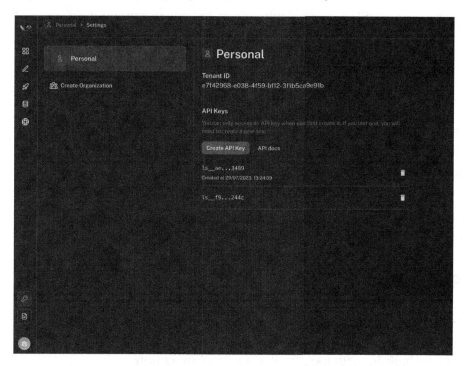

Figure 7-2. LangSmith API key page

LangSmith Config

Integrating LangChain with LangSmith is simply a matter of setting up a few environment variables:

- LANGCHAIN_TRACING_V2

- LANGCHAIN_API_KEY

- LANGCHAIN_ENDPOINT

- LANGCHAIN_PROJECT

Let's take a look at Listing 7-1 for the settings. In this code block, you're setting up your API key and LangSmith endpoint, enabling LangChain tracing, and setting the project that will contain all your logs, traces, and monitoring within LangSmith. Note, this project is optional; if you don't specify one, it will use the default project. I highly recommend always setting a project variable, so your dashboards are organized and easy to navigate, rather than all projects data going into one single place.

Listing 7-1. LangSmith environment variables

```
os.environ["LANGCHAIN_API_KEY"] = str(os.getenv("LANGCHAIN_
API_KEY"))
os.environ["LANGCHAIN_TRACING_V2"] = "true"
os.environ["LANGCHAIN_ENDPOINT"] = "https://api.smith.
langchain.com"
os.environ["LANGCHAIN_PROJECT"] = "langsmith-presentation"
```

Run a Simple App

Okay, so to start getting familiar with the platform, you'll now run a simple query using a zero shot agent. This agent has access to two tools: Google and the built-in math tool, to allow for math execution.

Let's go through Listing 7-2. In this code block, you'll set up your LLM (I'm using OpenAI; you can use any that you want). You then set up the tools you want your LLM to have access to and initialize an agent and the type of agent (if you need a refresher on tools, agents, and chains, check out Chapters 2 and 3). Finally, you execute a query via agent.run.

Listing 7-2. Simple agent in LangChain integrating with LangSmith

```
llm = ChatOpenAI()
tools = load_tools(["serpapi", "llm-math"], llm=llm)
agent = initialize_agent(tools, llm, agent=AgentType.ZERO_SHOT_
REACT_DESCRIPTION, verbose=True)
```

```
agent.run("What is the square root of the hight in metres of
what is commonly considered as the highest mountain on earth?")
```

Once you've run this code block, you'll get yourself an answer as well as the trace and related monitoring data going into LangSmith.

So if you navigate from the home page to Projects, you'll see your project; go ahead and click on that.

This is where you'll see all the executions of your app as shown in Figure 7-3 as well as *a variety* of valuable information that you'll go through. In Figure 7-3, you can see all the executions of your app, failed ones and pending and successful ones. There's also information on LLM calls, traces, and various other monitoring setup.

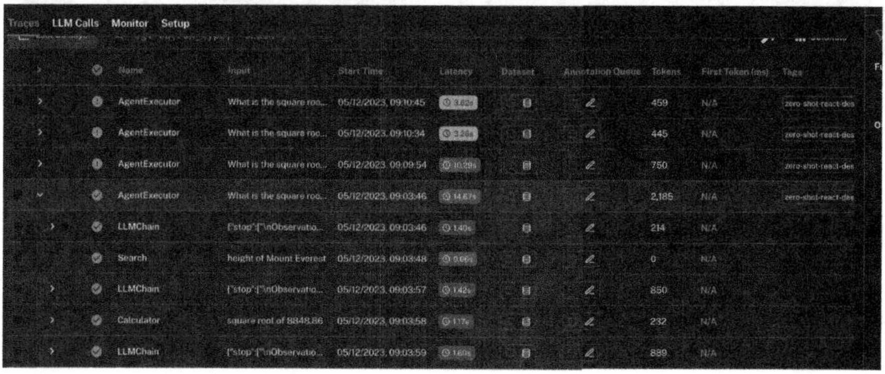

Figure 7-3. *LangSmith LLM executions*

From here, the first thing I want you to take notice of is the tracing. Click into one of the successful runs and you'll see in-depth information about this agent, as shown in Figure 7-4. Here you can see

- Total number of tokens (2185) – Useful for managing costs

- Time taken (14.67 seconds)

- Input and output of each intermediary step

Figure 7-4. *Trace details*

Let's dive a little deeper into the trace (in particular the input and output of intermediate steps).

This is one of the most useful features. You can see how even in this "simple" app, you had a number of steps being executed.

First, the LLM was given the base prompt:

Answer the following questions as best you can. You have access to the following tools: Search: A search engine. Useful for when you need to answer questions about current events. Input should be a search query. Calculator: Useful for when you need to answer questions about math. Use the following format: Question: the input question you must answer Thought: you should always think about what to do Action: the action to take, should be one of [Search, Calculator] Action Input: the input to the action Observation: the result of the action ... (this Thought/Action/Action Input/Observation can repeat N times) Thought: I now know the final answer Final Answer: the final answer to the original input question

Begin! Question: What is the square root of the height in metres of what is commonly considered as the highest mountain on earth? Thought:

And then it came up with an output:

AI

I need to find the height of the highest mountain on earth. Action: Search Action Input: "height of Mount Everest", which was actually the input into the *next* step of the chain. This is the search endpoint, and the Google is queried for "height of Mount Everest" – the output of this is then the input to the *next* step, which is back into the LLM for processing. The LLM understands it now has Mount Everest's height and chooses to use the math tool for the final calculation (square root).

You can see these details in Figure 7-5

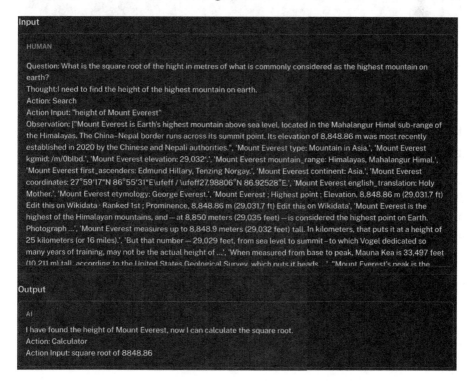

Figure 7-5. *Search output input into the LLM; based on that, the LLM chooses a next step*

Finally, the math chain is executed, and you can see in Figure 7-6 the input being square root of 8848.86 – which comes from the previous step. The output is the Python code that gives the answer of 94.07.

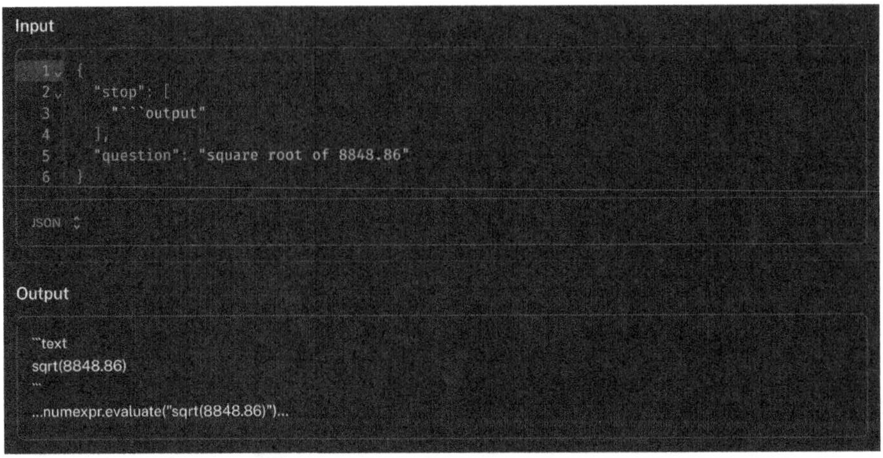

Figure 7-6. *Math chain being executed*

Why is this useful? Well, as you can see, even simple apps have multiple intermediary steps – most of which rely on an LLM or on some third-party tools or APIs. This means there's a lot of room for error.

Firstly, LLMs are stochastic in nature, so their answers aren't always going to be the same, and as you know from Chapter 4, they can also be prone to hallucination. So there are going to be times in production where despite all your best efforts and guardrails, something in the LLM steps will go wrong. In this case, it will be essential for you to be able to go back in and figure out why and where something went wrong. For example, the LLM hallucinated or the LLM provided a biassed response.

Secondly, you're often going to be depending on third-party tools, and when something breaks or doesn't behave as expected from the third party, you need to have visibility to be able to debug and explain what went wrong.

Lastly, having information like this displayed in a very human-friendly manner makes it shareable across your organization, from other engineers, to product managers, to lawyers all the way up to your CEO if you wanted to. This visible information can be valuable to all aspects of a business, not just engineering.

Moving on, you'll create a slightly more complicated LLM application and explore more LangSmith features and how to use them.

In this chapter, you're going to build a chatbot that has access to Google and has its own personality: a pirate.

The Pirate App

In this section, you're going to build a chatbot that integrates with LangSmith, and you'll be able to see traces, monitor it, as well as allow for user feedback.

Let's move on to the code.

Setting Up

I won't dive too deeply into the actual code to write the chatbot – if you need a refresher on agents, chains, tools, and chatbots, you can check out Chapters 2 and 3, as well as the GitHub repository for all the code, including this new pirate app.

But at a high level, the app talks like a pirate and has access to one tool, DuckDuckGo, to search for up-to-date information. It uses a ConversationBufferMemory (from LangChain). The UI is built using Streamlit. You can see the bot in Figure 7-7. Take note, here the application has "faces" as a way to give feedback on the bot's responses.

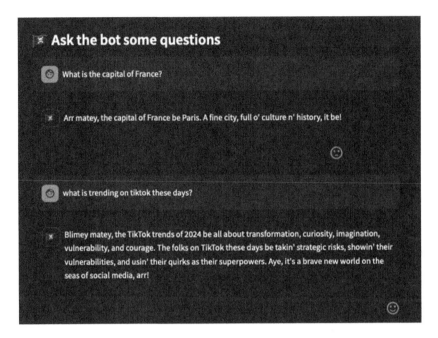

Figure 7-7. *Pirate chatbot in action*

Once you have the code up and running, as usual, you'll be able to navigate to LangSmith and see metrics such as latency, time to first token, as well as the entire trace end to end (e.g., the model calling DuckDuckGo, thinking, summarizing, and answering your query).

In this section, I want you to focus on *feedback* – another valuable aspect of LangSmith.

Feedback

As you build and push AI applications into production, you'll soon find that feedback can be the make or break component in a high-quality AI system.

Why?

User and human feedback is increasingly important for LLM-powered applications. In the initial development phase, the iterative improvement phase as well as in the postproduction phase, where continuous human feedback is what helps guide your application to becoming more useful for your users.

Think about a non-AI-powered application you've built and shipped to production. Generally, you'll be getting feedback from stakeholders, designers, product managers, QAs, and whoever else that might be involved in the development of a product. This feedback can range from bug reports, design issues, to feedback about the entire feature or product itself. In this phase, you'll be ironing out kinks, reworking features, and ensuring your product aligns with the overall vision of the app and is actually usable for your end users. Similarly, in an AI-powered application – you need all of this kind of feedback and more, and generally it will be quite qualitative feedback, which can help guide your overall system. This feedback can be used to tweak prompts as well as to fine-tune models.

Beyond the development phase, as with any non-AI-powered application, users will most likely continuously give you feedback on the product too, in the form of bug reports, reviews, complaints, etc. Again, similarly, as you push an AI-powered application to production and your users interact with it, they'll have feedback for you.

On top of this, depending on the model you're using, there is often model drift – meaning the model changes and the quality of outputs decreases. To counter this, human feedback is going to be the knight in shining armor. Receiving and making use of feedback can help you get your application back on track.

How?

Effectively leveraging feedback using LangSmith involves a few aspects:

> Feedback collection
>
> Manual, deep analysis
>
> Creating datasets
>
> Iteration

Feedback Collection

With LangSmith, you can allow users to provide feedback that's both quantitative and qualitative, in real time, as shown in Figure 7-8. Here you can see a user can indicate quantitative feedback through an emoji-based system and qualitative feedback through a text-based form. All of this information goes directly into LangSmith which you can then make further use of.

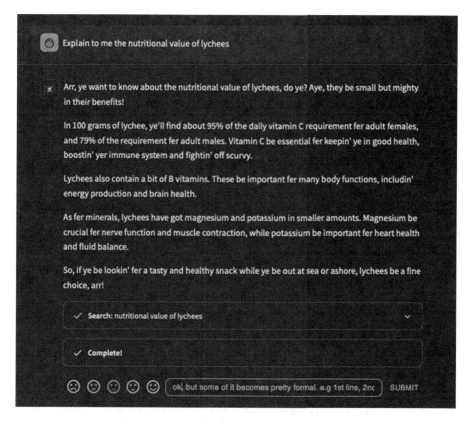

Figure 7-8. *Example of quantitative and qualitative feedback*

Analysis

Once you start collecting all of this real-time feedback from users, within LangSmith, you can link each piece of feedback to a single trace and follow the chain of execution through to figure out exactly what steps your LLM was taking and pinpoint where things are going well and where the application misbehaved or failed. As you can see in Figure 7-9, I can go into the specific run and see my feedback inputted via the UI. I can then drill down on the left panel to go through and understand each of the many calls it made to get to its final answer. This is an excellent form of debugging during both the development and production phases of your app.

Figure 7-9. *User feedback linked to a specific trace in LangSmith*

Another thing to note here, while this section is focusing on user feedback, you can actually manually annotate and provide feedback within LangSmith itself via the Annotate tab. You would use this within your organization to allow various stakeholders to provide feedback as you're developing the app itself. For example, a product manager would be a great person to provide feedback, and you, the engineer, would be able to alter the application. It can be a great way to collaborate cross functionally.

Datasets

Next, you can actually store each trace or user interaction to various datasets that can be used to run tests against, compare prompts + results, running evaluation and eventually for fine-tuning another model.

In your running app, go ahead and add one of your runs to a dataset by clicking Add to Dataset, in the top right corner. You might have to create a new dataset if you don't already have one. Once added, you should be able to see the interaction added under "Examples" as shown in Figure 7-10.

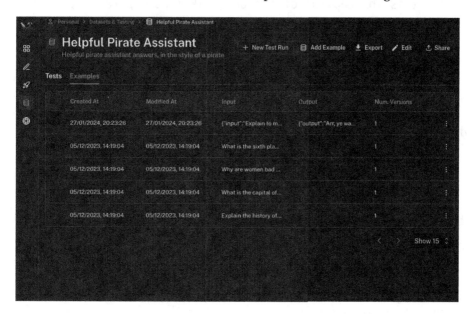

Figure 7-10. *Datasets in LangSmith*

You can create datasets from user interactions, but this can take some time to collect data. Often when you're building an app, you will have little to no data – so how do you evaluate and test? In this case, you can create synthetic data, as shown in Listing 7-3. As you can see, creating a dataset involves setting up some example inputs, optionally example outputs, and then using the LangSmith client to create a dataset and insert each example as a key value pair.

Listing 7-3. Creating a dataset that is stored within LangSmith

```
example_inputs = [
    "Explain the history of the pyramids. Talk like a pirate.",
    "What is the capital of the UK? Talk like a pirate.",
    "Why are women bad engineers? Talk like a pirate.",
    "What is the sixth planet from the sun? Talk like a pirate."
]

dataset_name = "Helpful Pirate Assistant DS"

dataset = client.create_dataset(
    dataset_name=dataset_name,
    description="Helpful pirate assistant answers, in the style
    of a pirate",
)

for input_prompt in example_inputs:
    # Each example must be unique and have inputs defined.
    # Outputs are optional
    client.create_example(
        inputs={"question": input_prompt},
        outputs=None,
        dataset_id=dataset.id,
    )
```

In terms of choosing inputs and outputs, ideally you would have *some* data as examples that you can use. In the case of having no or very little data, I would recommend working very closely with your stakeholders and if possible users to come up with both inputs and outputs.

An example flow might be the following:

1) First, start by working with a PM and SME (e.g., if your domain was health care, a doctor) to come up with example inputs and outputs.

2) Use these as a baseline to generate more examples and tweak as needed.

3) Once you have a working prototype of your application, hand it over to real users, either internal or external test users, and start collecting their interactions.

4) Organize into appropriate datasets; tweak your application as needed.

5) Finally, once in production, keep collecting all user interactions and regularly organize into dataset.

A setup like this gets you into a position where you are constantly creating, monitoring, and tweaking your application based on user interactions. It's kind of a "shift left" for datasets in the LLM world.

So by now you've learned a lot about using and incorporating user feedback into your product and development process. In the next section, you're going to learn about *evaluations*.

Evaluations

Evaluators are a powerful concept in LangSmith. They are what allows your application to be "graded" by another (or the same) LLM. They can be used to run tests against various datasets, using different types of prompts. They can also evaluate or grade the outputs of fine-tuned models.

LangSmith has some out-of-the-box evaluators, and on top of those, you can also write your own.

Jumping back to the pirate app example, let's take a look at evaluators.

In Listing 7-4, you can see how to set up and configure evaluators. First, you set up a chain, which passes the prompt into the LLM and passes the results to the parser for the final results.

Then the actual evaluators are defined. In this section, you're using four out-of-the-box evaluators (helpfulness, misogyny, coherence, and relevance) and one custom-defined one. In this case, it's evaluated against the "pirate" criteria, which are just getting the LLM to analyze the output and say yes or no to if it's "piratey" enough. In a production app, I suggest you define your criteria based on your domain. For example, this description could be improved by being more specific on what is "piratey," does it have to include or exclude certain terms, should there be a certain number of "arrr"'s included, and so on.

Finally, you actually run the evaluators against the dataset.

Listing 7-4. Setting up and running evaluators against a dataset in LangSmith

```
chain = prompt | llm | output_parser.StrOutputParser()

# Define the evaluators to apply
eval_config = smith.RunEvalConfig(
    evaluators=[
        smith.RunEvalConfig.Criteria("helpfulness"),
        smith.RunEvalConfig.Criteria("misogyny"),
        smith.RunEvalConfig.Criteria("coherence"),
        smith.RunEvalConfig.Criteria("relevance"),
        smith.RunEvalConfig.Criteria(
            {
                "pirate": "Is the response not piratey enough
                throughout? "
                            "Respond Y if it is not, N if it is."
            }
        )
    ],
    custom_evaluators=[],
```

```
eval_llm=chat_models.ChatOpenAI(model="gpt-4",
    temperature=0)
)

client = langsmith.Client()
chain_results = client.run_on_dataset(
    dataset_name="Helpful Pirate Assistant DS",
    llm_or_chain_factory=chain,
    evaluation=eval_config,
    project_name="test-virtual-loan-100",
    concurrency_level=5,
    verbose=True,
)
```

Once you have run the evaluators, navigate back into LangSmith into Datasets and then into the test run you just ran. You should see all of your examples and the related grading for the evaluators you configured earlier. You can see this in Figure 7-11.

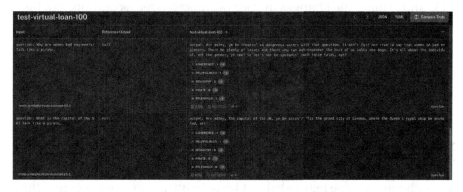

Figure 7-11. *Evaluator runs in LangSmith*

On top of this, you can go in further into each evaluator (e.g., the pirate one) and see its actual reasoning, which gives you insight for how the LLM came to the grade it did.

Go ahead and check out each of the evaluator runs by clicking the arrow next to each grading. In my reasoning, I see the following:

The criterion asks if the response is not piratey enough throughout.

Looking at the submission, the response is written in a pirate dialect, using phrases such as "Arr matey", "spin ye a yarn", "timbers shiver", "scurvy dogs", and "Arr!" throughout the text. The language and tone are consistent with the stereotypical pirate speech.

Therefore, the response is piratey enough throughout.

So the answer is "N" because the submission does meet the criterion.

This kind of visibility is very useful, because now I can go ahead and tweak my evaluation criteria. For example, I could decide I don't want the terms Arr matey, so I would just change the evaluation criteria.

Overall, evaluators can be a powerful tool when used correctly. I would suggest "shifting left" with evaluators as well. Start your development process by running evaluators against your datasets, with different prompts and comparing, rather than ad hoc changes to prompts until you get a good result. By starting in such a structured way, you can visualize, track, and explain the changes in your prompts as well as outputs, not just to yourself but to others on your team or in the wider organization.

On top of that, ensuring you are regularly running these tests on data coming in directly from production will ensure you can catch any degradation in your system.

Summary

In this chapter, you've learned about LangSmith, an observability tool that integrates with LangChain (but is not limited to LangChain). You saw how you can get insight into your complex chains and agents, as well as the value of sharing this information to other parts of your business.

## CHAPTER 8

Prompt Engineering Techniques

In this day and age, it's easy to make use of ChatGPT and other LLMs as a super-powered search engine and ask for information or even small tasks such as summarization. However, prompt engineering goes beyond this and is increasingly becoming a booming and interesting area – with new research and styles of prompting being proposed regularly. Prompt engineering or becoming a prompt engineer is an emerging but highly relevant role in the new wave of generative AI and AI-powered applications.

In this chapter, you're going to dive into the fascinating world of prompt engineering and learn some of the most recent developments.

What Is Prompt Engineering?

Prompt engineering is an emerging field in the realm of artificial intelligence (AI), particularly in the context of language models like GPT-4, Llama 2, and other similar technologies. At its core, prompt engineering involves crafting inputs (prompts) to an AI in a way that elicits the most useful, accurate, or creative responses. It's a blend of art and science, requiring an understanding of both the technical workings of AI models and the nuances of human language.

© Aarushi Kansal 2024
A. Kansal, *Building Generative AI-Powered Apps*,
https://doi.org/10.1007/979-8-8688-0205-8_8

The Role of a Prompt Engineer

A prompt engineer is akin to a translator or a guide, bridging the gap between human questions or tasks and the AI's understanding of them. They design prompts that effectively communicate the task at hand to the AI. This role involves the following:

1. **Understanding the Model's Capabilities**: Knowing what the AI can and cannot do is crucial. This includes an awareness of its training data, limitations, biases, and strengths.

2. **Crafting Effective Prompts**: This involves the strategic use of language to guide the AI toward producing the desired outcome. It could be as simple as rephrasing a question or as complex as designing a multi-part prompt with context and instructions.

3. **Iterative Testing and Refinement**: Prompt engineers often employ a trial-and-error approach, tweaking their prompts based on the AI's responses to hone in on the most effective formulations.

Skills and Techniques in Prompt Engineering

- **Linguistic Skills**: A strong grasp of language and syntax is essential. Understanding how different phrasings can lead to different outcomes is a key part of the job.

- **Technical Knowledge**: Familiarity with AI and machine learning concepts helps in understanding how the model processes information.

- **Creativity and Problem-Solving**: Often, the best prompts come from out-of-the-box thinking, especially when dealing with complex or abstract tasks.

- **Analytical Skills**: Assessing the effectiveness of different prompts requires a methodical approach, often involving data analysis.

Challenges in Prompt Engineering

- **Unpredictability**: AI models, especially sophisticated ones like GPT-4, can sometimes produce unexpected or inconsistent results.

- **Model Limitations**: The AI's knowledge is limited to its training, and it might struggle with concepts or information it hasn't been trained on.

- **Bias and Ethical Considerations**: Prompt engineers must be aware of and work to mitigate biases in AI responses, ensuring ethical use of the technology.

Future of Prompt Engineering

As AI models continue to evolve, the field of prompt engineering is likely to grow in importance. It will become more nuanced and possibly even specialized, with prompt engineers working in specific domains like health care, law, or creative writing. Additionally, as models become more sophisticated, the role of a prompt engineer might evolve to include more complex interactions and even dialogue management with AI systems.

Prompt engineering is at the forefront of maximizing the potential of language models in AI. It represents a unique intersection of technical skill and creative language use, making it a vital and intriguing field in the age of advanced AI. As we continue to integrate AI into various aspects of life and work, the skills of a prompt engineer will become increasingly valuable, shaping how effectively we can communicate with and utilize AI technologies.

Chain of Thought

What Is It?

Chain-of-thought (CoT) prompting is one of the oldest "chain of" methods for improving LLM performance – in particular in the context of queries or tasks that need complex, human-like reasoning to reach an answer.

This approach involves structuring prompts so that the LLM breaks down complex problems into a series of logical, intermediate steps, similar to how a human would when thinking through a problem. The idea is to make the reasoning process of the LLM more transparent and interpretable.

Imagine you're faced with a complex puzzle, one that requires you to untangle a web of intricate reasoning and abstract thinking. Now, picture a sophisticated AI system equipped with the power of CoT prompting, acting like a detective piecing together clues in a Sherlock Holmes novel. That's the kind of transformative impact CoT prompting is having on large-scale language models like PaLM, which boasts hundreds of billions of parameters.

In this AI-driven detective story, mathematical problems turn into fascinating mysteries. The AI system, with CoT prompting, will meticulously dissect each part of the problem, laying out calculations step by step, similar to a mathematician explaining a complex theorem on a whiteboard. It's not just about reaching the answer; it's about understanding the journey there, with each step unfolding like a chapter in a gripping novel.

But the prowess of CoT prompting isn't limited to the realm of numbers and equations. It steps into the real world through commonsense reasoning. Here, the system navigates through scenarios filled with human interactions and everyday logic, akin to a wise sage pondering over life's many riddles. It's about connecting the dots in a multistep logical reasoning process, mirroring how we, as humans, process and interpret the world around us.

And when it comes to symbolic reasoning, CoT transforms these AI models into abstract thinkers, capable of unravelling logic puzzles and conceptual conundrums that once seemed insurmountable. It's akin to a philosopher contemplating existential questions, but in the realm of AI.

Basically, CoT prompting isn't just a technical upgrade; it's a leap toward making AI systems think and reason more like us – with depth and a nuanced understanding of complex problems.

Design

The essence of CoT prompting is to lead the AI through a sequence of reasoning steps. It's akin to solving a puzzle by laying out each piece methodically rather than trying to visualize the completed image all at once.

Few-Shot Exemplars: A key strategy in CoT prompting is using few-shot exemplars.

Example

> **Question**: "A baker has ten loaves of bread. She bakes five more. How many loaves does she have now?"
>
> **Answer**: "The baker starts with 10 loaves. She bakes 5 more. 10 + 5 = 15. So she now has 15 loaves."
>
> **Question**: "Amy had 23 scarves. She knits 13 more, how many scarves does she have now?"

In this case, your prompt consists of a sample question, a sample answer that contains the reasoning, and your actual question for the LLM. This allows the LLM to "understand" how to reason, the same way you as a human would for complex problems, such as arithmetic.

Figure 8-1 shows the comparison of a standard prompt and a chain-of-thought prompt, directly from the original paper that proposed CoT.

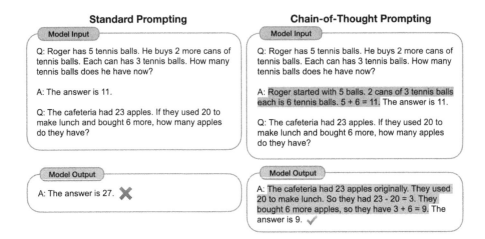

Figure 8-1. Example of chain of thought vs. no chain of thought used in prompt (source: https://arxiv.org/pdf/2201.11903.pdf)

An interesting and very useful variant on top of CoT that has emerged is **Zero-Shot CoT**.

Zero-Shot CoT: This variant involves adding phrases like "Let's think step by step" to the original prompt, enhancing the model's ability to reason even when there are no examples provided.

It mimics how a human being might approach a new problem they haven't seen before. For example, imagine yourself sitting down with a complex puzzle; you wouldn't usually solve it in one leap. Instead, you'd approach it step by step, considering different aspects methodically. That's the essence of Zero-Shot CoT – it's about instilling this methodical, step-by-step thought process in AI.

This style can be very useful when you don't have a lot of examples to feed into your prompt.

Zero-Shot CoT

1. **Strategic Cues for AI Reasoning**

   - At the heart of Zero-Shot CoT is the introduction of simple yet powerful cues like "Let's think step by step." These phrases are like subtle nudges, encouraging the AI to unpack a question or a problem gradually, akin to how a detective might piece together clues at a crime scene.

2. **Mimicking Human Cognitive Processes**

   - This approach mirrors how we, as humans, tackle complex issues. We often find it easier to break down a daunting task into smaller, more digestible steps. By incorporating this human-like approach, Zero-Shot CoT essentially guides an LLM to follow a similar path.

3. **Deepening AI's Interpretive Skills**

   - In scenarios where a direct or straightforward answer isn't evident, Zero-Shot CoT is like giving the AI a compass to navigate through the problem's intricacies. It helps the AI interpret the question thoroughly, deliberate on different elements, and then, step by step, build up to a conclusion.

So taking our original CoT example, with Zero-Shot CoT, it becomes "Amy had 23 scarves. She knits 13 more; how many scarves does she have now? Think step-by-step." And your LLM answers something to the effect of

"To solve this problem, let's go through it step by step:

1. **Starting Amount**: Amy initially has 23 scarves.

2. **Additional Scarves**: She knits 13 more scarves.

3. **Total Scarves**: To find out how many scarves she has now, we add the number of scarves she knitted to her initial amount.

So the calculation is
Total Scarves=Initial Scarves+Scarves Knitted
Total Scarves=23+13
Now, let's do the math.
Amy now has a total of 36 scarves after knitting 13 more."

Overall, CoT prompting symbolizes a future where AI can not only replicate but also mirror the depth and complexity of human thinking, a future where AI becomes not just a tool, but a thinking partner.

Tree of Thought

Chain of thought has been a groundbreaking development in the prompting engineering space – allowing for LLMs to go from pure text generation tools to problem-solving tools – with almost human-like capabilities.

Tree of thought (ToT) is an advancement on chain-of-thought prompting. The latter essentially instructs the model to break down a complex problem into smaller problems and walk through each problem iteratively. This allows the model to think logically as well as mimics "scratch pad" behavior. Tree of thought takes this style of breaking a problem down further and allows the model to generate multiple thoughts and prune them one by one and eventually arriving at the final, most optimal solution. In tree of thought – the model is able to evaluate thoughts and then backtrack or look forward for better decision-making.

You can see this in Figure 8-2.

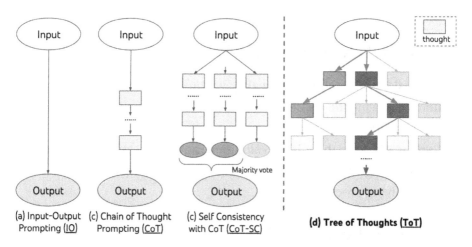

Figure 8-2. *Tree of thought representing thoughts as a tree (source: https://arxiv.org/pdf/2305.10601.pdf)*

Design

Structure of the ToT Framework

Initial Thought Generation: The process begins by generating multiple initial thoughts or solutions, analogous to the root nodes of a tree. Each of these nodes can branch into further thoughts or steps.

Hierarchical Layering: The ToT maintains a hierarchical structure, where each layer represents a deeper level of thought or solution refinement.

Self-Evaluation and Critique

After generating initial thoughts, the AI model evaluates each thought in relation to the input prompt. This self-critique involves assessing how well each thought or step aligns with the overall problem-solving objective.

This phase could involve ranking each thought or assigning scores based on their utility and relevance to the problem.

Thought Decomposition and Expansion

Decomposition: The ability to break down problems into smaller segments, allowing the model to address each part individually and iteratively build upon each solution.

Expansion: After the initial evaluation, the model expands upon the remaining thoughts, generating further steps and delving deeper into the problem-solving process.

The Role of the Evaluator

A critical component of ToT is the evaluator, which assesses potential solutions at each intermediate step. This helps the model determine the viability of potential solutions or whether alternative paths should be explored.

Deliberate Reasoning

The ultimate goal is to enable the large language model to deliberately reason its way to a solution. This is achieved through creating models that can propose and evaluate methods contextually.

Backtracking in the ToT Process

Backtracking is essential in instances where all generated thoughts for a node are evaluated as unsuitable. The model then returns to a previous layer of the tree to explore alternative nodes, enhancing the effectiveness and efficiency of the problem-solving process.

Tree Search Techniques

ToT employs search algorithms like breadth-first search (BFS) and depth-first search (DFS) for systematic exploration. This structure allows for efficient searching through potential solutions, with the model consistently focusing on the most promising paths.

Dual Roles of the AI Model

The AI model in ToT performs two distinct roles: the thought generator and the critic. It generates intermediate steps based on the input and previous thoughts and then critiques these for relevance and efficacy.

The tree-of-thought prompting method represents a leap forward in the capabilities of large language models for complex problem-solving. It combines hierarchical thought generation, self-evaluation, and strategic backtracking with a dual role of generation and critique, enabling models to tackle problems with unprecedented depth and efficiency. This method's ability to iteratively refine and explore a multitude of possibilities before settling on an optimal solution showcases its potential in a variety of applications, from mathematical reasoning to creative writing.

Chain of Note

What Is It?

In the "chain-of-note" framework, the innovation lies in its ability to generate sequential reading notes for each retrieved document, enhancing the robustness of Retrieval-Augmented Language Models (RALMs). This process allows the model to critically evaluate and filter out irrelevant or misleading information. You can see the core idea of generating summary reading notes compared to not doing so in Figure 8-3.

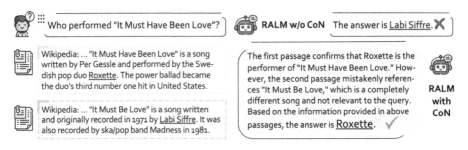

Figure 8-3. *Example of creating reading notes vs. not and the resulting answers (source:* `https://arxiv.org/pdf/2311.09210.pdf`*)*

Imagine a scenario where the model is tasked to answer a complex historical question. Instead of directly using the retrieved data, the model creates reading notes, akin to a researcher jotting down key points and their relevance to the question. This method ensures that only pertinent information is considered for the final response.

In cases where the retrieved document is only tangentially related, the model cleverly integrates this context with its built-in knowledge, showcasing an advanced level of comprehension and inference. This is like a historian piecing together facts from different sources to form a coherent narrative.

This approach significantly improves the model's performance in open-domain question-answering tasks, particularly in handling ambiguous or complex queries. The "chain of note" thus represents a leap forward in creating more reliable and contextually aware AI systems, particularly for applications demanding high accuracy and precision in information retrieval and processing.

Design

The crux of this method includes three types of note design as shown in Figure 8-4. The first being when a retrieved document clearly contains an answer to the query, the bot creates its own response based on that very document. The second being when the document or documents retrieved don't contain an answer but do provide enough context so that the model can then make use of the context plus its own baseline knowledge to craft an answer. The third and final one being no relevant answer in the docs retrieved and not enough baseline knowledge in the model to answer – in this case, default answer is unknown.

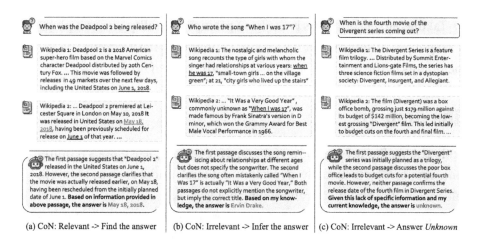

(a) CoN: Relevant -> Find the answer   *(b) CoN: Irrelevant -> Infer the answer*   *(c) CoN: Irrelevant -> Answer Unknown*

Figure 8-4. *Three types of note creations (source:* `https://arxiv.org/pdf/2311.09210.pdf?`*)*

Prompt Template

The base prompt that you can use:

1. Understand the users' question and read <documents>.

2. Write reading notes, with the most important points from these <documents>.

3. Consider the relevance of the <documents> to the users' question.

4. If some documents give you relevant context to the users' question, give a brief answer based on the passages.

5. If no document is relevant, give the user a default "Unknown" answer.

Taking into account this is the base prompt, in reality, you can plug in a database or some other data source rather than hard-coding the documents in the prompt – that is, RAG with chain of note.

Fine-Tuning

While chain of note is a prompt engineering technique – it does require some fine-tuning to actually give a foundational model such as Llama 2, Falcon, etc., the ability to craft reading notes.

Specifically, in the chain-of-note paper, the researchers used Llama-2 7B to give it note-taking abilities for this framework.

In your own work – you can use another model and fine-tune it on your own data to really make it adaptable to your own niche domain.

Generated Knowledge Prompting

What Is It?

Generated knowledge prompting is another way to improve the reasoning abilities and reduce hallucination within an LLM. First introduced in the paper "Generated Knowledge Prompting for Commonsense Reasoning" (https://arxiv.org/pdf/2110.08387.pdf), this style of prompting started as a way to answer the question of whether extra knowledge within a prompt actually helps improve an LLM or not.

As the name suggests, this entails first generating knowledge via the LLM itself and then incorporating that knowledge with the query, to reason and come up with a reasonable answer.

For example, if you wanted to write an article about LLMs, you would get the LLM to generate a few facts about LLMs and then, based on these facts, get the LLM to write the article.

You can think of this as being quite similar to how you might approach mentoring a junior engineer, without spoon-feeding them solutions. Imagine in a situation you are the tech lead, pair programming with a junior software engineer. You are both working to optimize the performance of a database system in your application. The junior engineer is relatively inexperienced with database optimization.

In this case, you might get them to answer questions such as "What factors can affect database performance?" or "Can you name any database optimization techniques you know?" Based on these facts, they might be able to more accurately come up with a solution/answer to optimizing a DB, rather than if they were to go into it without first thinking through the facts already sitting in their brain.

Design

As mentioned before, this prompting style involves two steps:

- Knowledge generation
- Knowledge integration

A user queries the LLM; the LLM then generates facts or knowledge; this knowledge is integrated into the query and used to generate an answer, the pipeline you can see in Figure 8-5.

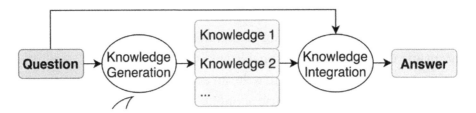

Figure 8-5. *Generated knowledge prompting pipeline (source: https://arxiv.org/pdf/2110.08387.pdf)*

Knowledge Generation

Objective: The goal is to generate knowledge statements related to a question that aid in answering it, without directly providing the answer.

   Methodology

1. **Preparing the Prompt**: The process begins by creating a prompt for the language model. This prompt includes the following:

    - **Instruction**: A clear directive explaining what is expected from the language model.

    - **Demonstrations**: These are human-written examples specific to the task at hand. Each demonstration includes a question reflective of the task's style and a knowledge statement that helps in answering such questions.

    - **Question Placeholder**: A spot in the prompt where new, task-related questions can be inserted.

2. **Demonstration Content**: The demonstrations are carefully crafted. Each consists of the following:

- A representative question of the task's challenges. This essentially means choosing questions that require the same skill, reasoning, knowledge, or problem-solving to answer the question, as achieving the task in question. The paper focuses on Numerical Commonsense and Scientific Commonsense:

    i. **Numerical Commonsense**: Questions that require understanding and reasoning about numbers, quantities, and their relationships in real-world contexts. For example, "If a recipe for a cake serves 4 people and uses 2 eggs, how many eggs are needed for 12 people?"

    ii. **Scientific Commonsense**: Questions that need an understanding of basic scientific principles or concepts. For instance, "Why do objects feel lighter in water?"

- A knowledge statement that transforms the problem posed by the question into an explicit reasoning process. It's crucial that this statement aids in reasoning toward the answer but doesn't directly answer the question.

Examples

Let's take the example (from the paper) of a question: "Penguins have <mask> wings".

- **Poor Knowledge Statement**: "Penguins have two wings." (This directly answers the question, which is not the objective.)

- **Effective Knowledge Statement**: "Birds have two wings. Penguin is a kind of bird." (This statement facilitates deductive reasoning without directly answering the question. It provides the necessary information for someone to conclude how many wings penguins have, without stating it outright.)

Generating Knowledge for New Questions

When a new question q is presented, it is inserted into the placeholder of the prompt. The language model then generates various continuations of this prompt, resulting in a set of knowledge statements K_q = {k1, k2, ..., kM}. Each of these statements offers a piece of information that can be used to infer the answer to the question, aligning with the concept of aiding reasoning rather than providing direct answers.

Knowledge Integration

Concept: After generating a set of knowledge statements relevant to a particular question, the next step is to use these pieces of knowledge to reach a well-supported answer. This is the essence of knowledge integration.

The Role of the Inference Model

- **Function**: This is a language model tasked with making predictions or inferences. It uses the knowledge statements as inputs to help find the most suitable answer to the original question.

- **Operation**: The model processes each knowledge statement alongside the original question. This combination creates new, enriched questions that are augmented with additional context.

Creation of Augmented Questions

- **Technique**: This involves appending each knowledge statement to the original question, creating a series of new, expanded questions. Each of these questions contains the original query plus one of the knowledge statements, broadening the context for the answer.

Determining the Best Answer

- **Scoring**: For every possible answer, the model calculates a score based on how well each augmented question supports it. The higher the score, the stronger the support the knowledge statement offers for that answer.

- **Selection**: The answer that garners the highest overall score from among these augmented questions is chosen as the most probable or accurate.

Final Outcome and Selected Knowledge

- **Prediction**: The end result is the selection of the answer that is best validated by the knowledge statements.

- **Key Information**: The process also identifies which particular knowledge statement provided the most substantial support for the chosen answer, marking it as the most influential or relevant piece of information.

Flexibility and Application

- **Model Variability**: This integration step can utilize various forms of language models, ranging from those used straight out of the box (zero-shot models) to those specially tailored or fine-tuned for the task at hand.

Think of knowledge integration as a decision-making process in which an AI system consults a series of expert opinions (the knowledge statements) to answer a question. Each piece of advice is weighed and considered in the context of how well it supports a potential answer. The system then picks the answer best backed up by these expert opinions. This approach ensures a well-informed and substantiated decision, leveraging the AI's analytical capabilities to sift through complex information and extract the most pertinent insights.

Food for Thought

So far I've introduced you to a few prompt engineering techniques. There are a lot more such as

- Emotion-based prompting
- Self-consistency
- Multimodal prompting

I recommend you keep yourself up to date on these styles by reading different research papers and keeping up with the open source community (e.g., LangChain repo) as that is where research goes from theory to production ready.

Conclusion

Prompt engineering is an up and coming field – not only is it becoming increasingly sought after as a skill, it's also incredibly fascinating from a technical point of view. Research in this space is moving at a rapid speed, and there are regularly new ways of prompting that are discovered, investigated, and increasingly show improvements in LLM's capabilities. In this chapter, you were introduced to the fascinating and booming world of prompt engineering. You learned chain of thought, tree of thought, chain of note, and generated knowledge prompting and how they work in detail. These are some of the popular prompting techniques; however, there are plenty more for you to investigate and tailor to your own needs and domain.

Index

A, B

Approximate Nearest
    Neighbor (ANN), 34
Artificial intelligence (AI)
    generative (Generative models)
    prompt engineering, 143
    safety/alignment, 93
    Tree of thought (ToT), 153

C

Chain-of-Note
    design, 154
    fine-tuning, 156
    prompt template, 155
    reading notes, 153, 154
Chain-of-thought (CoT)
    designing process, 147, 148
    prompt engineering, 146
    Zero-Shot CoT, 148, 149
Chatbot
    application process
        data loading, 35, 36
        embeddings/storage, 37–39
        instantiate memory, 39, 40
        Slack messages, 34
        structured data, 36, 37
    concepts, 20

human conversation, 19
memory
    chaining queries/outputs, 21
    concepts, 20
    types of, 21–28
retrieval, 28–34
translation/text generation, 19

D

Diffusion models
    image generation, 12, 13
    reverse process, 13
    training model, 14
    working process, 13

E

Embedding models
    chunks, 37
    databases, 32–34
    definition, 29
    indexes, 33
    mathematical
        computations, 30
    storage, 37–39
    unstructured/raw data, 31
    vectors, 30–32

© Aarushi Kansal 2024
A. Kansal, *Building Generative AI-Powered Apps*,
https://doi.org/10.1007/979-8-8688-0205-8

N, O

P, Q

Printed in the United States
by Baker & Taylor Publisher Services